BEAUTY
and the
BEAST

BEAUTY and the BEAST

The Coevolution of Plants and Animals

SUSAN GRANT

Illustrations by
LASZLO KUBINYI

CHARLES SCRIBNER'S SONS • NEW YORK

*To Professor Howard Bigelow, of the
Botany Department of the University
of Massachusetts at Amherst,
with thanks for introducing me
to the concept of coevolution.*

Copyright © 1984 Susan Grant

Illustrations © 1984 Laszlo Kubinyi

Library of Congress Cataloging in Publication Data

*Grant, Susan.
  Beauty and the beast.*

  *Bibliography: p.
  Includes index.
  1. Coevolution.   I. Title.*
QH372.G73   1984   575   84-10609
ISBN 0-684-18186-X

CONTENTS

Contents

Many people helped me while I was working on this book. The most important is my husband Don, whose encouragement helped me get over many rough spots, and whose critical eye detected a number of problems in the work that I was (ultimately) happy to correct.

My graduate school advisor, Ted Sargent, taught me how to search the biological literature and present facts gathered from it in an orderly way. He and my other guidance committee members, Howard Bigelow and Herb Potswald (now deceased), helped build my confidence in my ability to create a book-length manuscript.

The library staff at Holyoke Community College, especially Ginny Malone and Bob Stoddard, helped me find necessary references and let me keep them long enough to actually use them. The library staff at Smith College's Science Library—Pam Walton, Chi-Ping Zhang, Klara Dienes, and Gail Adametz—did the same thing. They also provided me with a nice writing space away from home.

My student Gail Thompson suggested an interesting theory about birds and beak-marked butterflies, which is mentioned in Chapter 5 for some enterprising researcher to pursue. Bill Hutchinson, Chairman of the Division of Biological Sciences at Holyoke Community College, shared with me the Chinese proverb on longevity mentioned in Chapter 9. He also shared his enthusiasm for this project and his interest in plant ecology.

The typing of the manuscript was carried out with cheerfulness and precision by Sandy Reed. And last but not least, Lucie Swain provided high-quality child care for my daughter, April Catherine Grant, during most of the time I worked on this book.

CHAPTER I

Mutual Injury, Mutual Aid: Patterns in Coevolution

THE LIVING-TOGETHER ARRANGEMENTS OF ANIMALS and plants are incredibly varied—all the way from instances of mortal foes to mutual benefactors. A gypsy moth caterpillar that devours young oak leaves, only to have its offspring killed later by poisons in the older leaves, is a good example of mutual injury. A flower-feeding bat that licks nectar from cactus flowers, and cross-pollinates them in the process, is just as good an example of mutual aid. How did relationships like these begin, and how do they influence the lives of the creatures involved?

Our answer to the question of origins was once that animals, plants, and their relationships were created by God. This explanation, widely accepted until the late nineteenth century, has had another surge of popularity through the efforts of creationist writers. Some animal–plant relationships—the termites that raise their own mushrooms, the trees that keep a standing bodyguard of ants—look so beautifully designed it is hard, at first, to reject the idea of a designer. But both of these relationships are mutualisms: plant and animal participants share in the benefits. When we turn to what might be called "the dark side of the Force" and encounter poisonous, bitter, stinging plants devoured by voracious animals, it becomes harder to accept such antagonisms as the work of a benevolent deity.

Current biological thinking sees mutualisms and antagonisms alike as the outcome of chance association, fine-tuned by evolution. In that light, the burning stings of a nettle plant and the blistering oil of poison ivy make sense, not as the handiwork of a malicious creator, but as biological weapons invented by accident, and honed sharp by generations of natural selection. The nectar that a honeybee laps up makes sense too, as a bribe invented by chance but improved in seductiveness over the ages to attract and feed a pollinator. Accident can generate an association; over the years, any inheritable change for the better in the plant or animal involved will be favored by natural selection. If it appears that an animal–plant association has promoted the evolution of special adaptations in its participants, that is coevolution in action.

Coevolved antagonisms, obvious and numerous, have tempted many to picture these interactions as unhappy marriages, complete with sex-role stereotypes. The animal participant is seen as an aggressive, destructive male, the plant participant as a passive, submissive female—a flowering Beauty condemned to devastation by a Beast. William Blake's couplet "The caterpillar on the leaf/ Reminds thee of thy mother's grief" is an elegant crystal-lization of this attitude. But it is unfair to the animals and plants involved, and even unfair to the original Beauty. In the fairy tale, Beauty herself chooses to live with the Beast in order to save her father, but at first she rejects the Beast's proposals of marriage. When she does finally accept him, it is her own decision.

Many if not all wild plants attacked by herbivores are like Beauty in the sense that they are taking the risks for which they are best suited. They are bound to be exposed to certain types of animal attack because of the life histories they have evolved. And like her, most of them manifest quiet stratagems of resistance to their foes. Many can control the growth and reproduction of their animal antagonists, who enjoy at most a Pyrrhic victory over the plant world. (This is not the case with cultivated plants, however. The result of long years of human selection rather than natural selection, they tend to score high on tastiness, succulence, and productivity, but low on self-defense abilities. Like the aristocratic girls of ancient China whose feet were bound and crushed in childhood, cultivars are for the most part helpless dependents on the society that made them what they are.)

Coevolved mutualisms have attracted more lyrical description than antagonisms. Writers on mutualisms generally emphasize the sense of rightness, of compatability that these relationships give us. Yet in many cases these apparently perfect partnerships have not only grown out of accidental associations but out of antagonistic roots. Beetles, the hypothesized earliest pollinators, still visit flowers like magnolia, gnawing on their petals. This damage has evolved over the ages into a source of genetic vigor: the beetles pick up pollen in the process and transfer some of it to other magnolia flowers. Such primitive pollination strategies as these linger on, but there now exists the more intense mutualism of bee-plant pollination. Bees need food from flowers for daily survival; bee-flowers need their bees to set seed. Just as Beauty finally recognized that she loved her Beast, many plants involved in mutualisms have evolved a need for their animal partners. Interdependence has replaced original neutrality or injury.

How large a part does any coevolved relationship play in the daily life and ecological success of its participants? The answer to this question is as yet unknown, in all but a few cases. Coevolutionary study is a young branch on the tree of knowledge. Even though some pollination partnerships, for example, were already being examined in Darwin's day, new ones are still turning up. Few coevolved relationships have yet been subjected to the kind of rigorous quantitative analysis that would measure their exact impact on their participants. The roles of plants and animals in seed dispersal are among the exceptions. The results confirm our intuitive suspicion that animals' impact can be great on the transport and survival of seeds. And in cases where marine animals harbor tiny mutualistic green plants, careful studies have found that these algae play an immense role in feeding their animal hosts.

Occasionally coevolutionary assumptions have had to be reversed as the result of new studies. The now-classic example is the role of ants in relationship to flowering plants. Notorious sugar-eaters, ants have been viewed for years as mere nectar thieves, depriving legitimate pollinators of their reward. But field studies in the tropics, to be discussed in Chapter 2, show that many plants have evolved the ability to attract ants with nectar glands or food pellets on stem and leaf. The ants benefit such plants: lured by these enticements, they defend the plants against other, more destructive herbivores. They bite and sting any would-be consumers of their plant. Thus the role assigned to ants in coevolutionary drama is currently being revised in cases such as these, from villain to hero.

The questions that coevolutionary theory prompts us to ask are testable ones. Ideally, a question like "What advantage does milkweed gain from being poisonous?" should be answerable by research in the natural environment. Many of the studies discussed in this book were carried out in a natural setting. But laboratory work may also be essential in order to find out just which animals the plant can affect, and how survivors later react to the plant. In a similar vein, when we ask what good foxglove derives from having speckled flowers, both field and laboratory research play a part in deciphering the cryptic symbols these flowers display.

An aspect of coevolutionary study that makes it particularly challenging is the fact that coevolved relationships are always vulnerable to change. Sometimes this change may be for the worse, as when plants dependent on hummingbirds for pollination are transported to islands without them. Under such circumstances, if the plants happen to acquire new partners in pollination, all is well, but the relationship must be redefined by the biologist. In other cases, the change may involve a gain for the species. When a herbivore like the gypsy moth travels (courtesy of human hands) from its homeland to new territories, its status may alter from being a minor pest on plants well-adapted to its attack, to being a major pest on new victims. As time passes, the picture may change again, due to the slow guerrilla warfare that the new food plants begin to wage against the invader.

In addition, even a slight advantageous change in the plant or animal that is inheritable, due to mutation, will help insure a lead in the struggle for survival. And mutations can occur anywhere, at any time. Even long-established animal–plant relationships never disrupted by natural or forced emigration are still open to alteration, as spontaneous genetic changes occur and are winnowed out by natural selection.

A wide range of specialized training is called into play in the study of coevolved relationships. Amateurs and professionals in animal behavior, entomology, plant and animal ecology, horticulture, and even biochemistry, have all become involved in research on both antagonisms and mutualisms. A beautiful aspect of the field which its students share is an awareness of the many questions that still need to be answered. In the chapters that follow I will explore the major categories of antagonisms and mutualisms, and point out the unsolved puzzles that they hold.

Self-Defense: Barbed Wire and Bodyguards

ANTAGONISMS BETWEEN PLANTS AND ANIMALS HAVE often been compared to human warfare. The analogy is an apt one in several ways. In human conflict, a superpower that is prepared for war against a similarly powerful enemy may prove surprisingly vulnerable against an unfamiliar foe using unconventional methods. Leonard Wibberley's novel *The Mouse that Roared* pictured such a scenario—a handful of soldiers wielding longbows succeed in seizing control of Washington, D.C. A more tragic real-life parallel occurred during the Vietnam War, when the United States with its huge armory was unable to defeat guerrilla resistance in Southeast Asia. In many coevolved antagonisms a plant survives as long as it faces only familiar foes. A new antagonist may devastate it.

The development of novel weapons by a human government prompts the counterdevelopment of a new arsenal by the enemy. We often see such an arms race in progress in a coevolved antagonism. Gypsy moth caterpillars and oak trees, as mentioned earlier, are a good example. Oaks secrete chemicals that deter many animals from eating their leaves, but that fail at first to repel gypsy moth larvae. The caterpillars devour oak leaves in New England in May and June. Later on that summer, or the next summer, the oaks produce new leaves measurably more poisonous than their first batch. The caterpillars—hooked on oak by now—keep right on eating. Finally most of them give up the ghost—and the tiny pellets of caterpillar excrement stop dropping from the trees. A few genetically tougher caterpillars survive to perpetuate the species—and the antagonism.

Plants that evolve on islands are especially vulnerable to animal destruction because of their long isolation from such attack. When rats and goats come ashore from visiting ships, most island plants might be compared to the early English monasteries hit by Viking raids. The native wild plants on the Galápagos Islands were drastically damaged by goats, and were saved only by a goat-free sanctuary created by human beings. Native Hawaiian plants (and some of the native animals) have been hit even harder by invading rats, with many of the aboriginal species now eradicated.

One more parallel between human warfare and coevolved antagonisms lies in the vulnerability of each side to changes in the other's strategy. In human warfare, such unpleasant surprises are forestalled by spying. In nature, if an unexpected mutation raises the resistance of a plant, the animals that feed on it are usually caught off guard. Their only chance for survival then lies in shifting to a new food plant, or else in the very slim possibility that a comparable genetic change will let them continue to exploit their original victim.

A widespread, simple, and remarkably effective strategy for plant self-defense is what might be called the barbed-wire syndrome: the possession of bristles, hooks, spines, and thorns. Desert plants, from cacti to crown of thorns, have long exhibited such weaponry, and in doing so many of them illustrate a phenomenon known as convergent evolution. Cacti and crown-of-thorns plants are unrelated. This becomes obvious when they flower; the tiny blossoms on a crown of thorns are nothing like the deep trumpets of cacti. But the fat, thorny stems look remarkably similar. Exposed to the same kind of pressures from dry (and hungry) environments, they have evolved similar protective devices.

Holly trees and thistles also exhibit the barbed-wire syndrome. Leaf-eating caterpillars will devour a holly leaf with its spines cut off, but refuse to take a nibble when the spines are left on. And thistles—even more heavily armored—stand up to almost all their enemies. The beautiful purple flowers of the thistles are often the only ones visible in a heavily grazed pasture. The defiant strength of thistles has made them the chosen emblem of Scotland.

But even they have a few successful enemies: caterpillars of painted lady butterflies feed comfortably on thistle leaves, no matter how spiny.

The leaves of many plants are covered with tiny hairs or trichomes that act as physical defenses on a smaller scale. Nettle-leaf trichomes have swollen tips that release droplets of irritating fluid when touched. Trichomes on cultivated geranium leaves have similar-looking tips which release aromatic oils upon contact. Pleasant to smell but unpleasant to taste, the oils repel many leaf-eating insects. Certain varieties of bean plant grow trichomes tipped with hooks on their leaves. The hooked trichomes impale a major pest of beans, the leaf-hopper, many of which bleed to death after being wounded. A few hoppers manage to shed their hooks by shedding their old skin as they molt to a larger stage. It's tempting to imagine that trichomes might come in handy in that one situation, rather like a bootjack that helps remove a tight pair of boots. The trichomes on bean plants are most abundant on the youngest, tenderest leaves, the ones most attractive to leaf-hoppers. In this battle, the beans have concentrated their defenses where they will do the most good.

Certain types of passionflower vines, common in the tropics, grow similar trichomes on their leaves. When caterpillars of the *Heliconius* butterfly, a feeder on passionflower leaves, are placed on such bristly plants, the trichomes wound their soft skins. They bleed to death or starve to death, pinned down by tiny trichomes like Gulliver among the Lilliputians. But why have only some species of passionflower come up with this defense? Trichomes may be a recent innovation in the family; other species of passionflower rely on poison in their smooth-surfaced leaves.

Caterpillars of the *Mechanitis* butterfly feed on plants in the potato family, and thrive on them in spite of the fact that many of these plants also have leaves that bristle with trichomes. How do the larvae avoid being impaled? At least one species of *Mechanitis* has evolved a group response to the problem. The caterpillars gang up to spin a thick silk web that covers the trichomes, and then crawl over the leaf on this safety net. Their sociability, unusual in *Mechanitis* species, thus creates an effective response to the plant's spiny defenses.

A different, more subtle defense—one based on deception—protects other passionflower vines. They grow little yellow bumps and projections on the tips and near the bases of their leaves. Butterflies, especially *Heliconius* butterflies, are strangely reluctant to lay their eggs on such vines. Insect behavior specialists believe that the female butterflies mistake these little yellow bumps for eggs of their own species. In a laboratory study, *Heliconius* females were unwilling to lay eggs on a vine occupied with real eggs, and just as unwilling to lay on a vine armed with "mimic eggs."

Why should they be so reluctant? A partial answer has emerged from observations on similar behavior in the common cabbage white butterfly. Females of this species also refrain from leaving eggs on already occupied plants. This makes sense biologically: a crowded food plant will be a poor larder for the larvae. A female butterfly's ability to discriminate against such occupied plants is one of the greatest gifts she can possess; she thereby gives her offspring the food they need to survive. Swallowtail butterflies show this avoidance as well as the cabbage whites. In one study, a female swallowtail that had almost settled on an appropriate plant flew away abruptly when she detected eggs on it.

In all these cases, butterfly avoidance of a truly or deceptively occupied plant is a good thing for the plant. In at least one other case, however, similar reluctance benefits only the insect, which then exploits the plant quite thoroughly. This situation occurs with the apple maggot fly. Adult females of this species mark the surface of an apple with a scented secretion, once they lay an egg in it. The scent conveys a message, roughly translated as "the apartment's been rented," to other female apple maggot flies. They avoid laying eggs in the occupied fruit, which is reserved for just one larva. If the message-bearing secretion or pheromone can be extracted and used on a commercial scale, it may prove a boon to apple-growers.

Pods, husks, and shells provide another line of plant self-defense, a line that has been especially well-studied in two western trees: the lodgepole pine and Douglas fir. Squirrels love to nibble the seeds from cones of both these trees, but they prefer the Douglas fir because it produces big crops of seed-filled cones that are easier to open than the pine's. But the fir is a sporadic producer. Some years are good crop years for Douglas fir but others are lean indeed, with no seeds produced at all.

When the fir-cone crop is poor or nonexistent, squirrels

turn to the lodgepole pine cones. But these are a tough nut to crack. In the dry part of their range, the area just east of the Pacific Coast Cascade mountains, the lodgepole pine grows cones with dense, heavy scales that refuse to open spontaneously. They stay closed unless seared by fire, so that they may remain tightly shut for years.

The squirrels that live in that range have adapted to those cones. They are bigger, stronger, and have heavier jaw muscles than those in the damper West Coast forests where softer cones abound. In one study, careful measurement of squirrel skulls preserved in a western museum revealed that almost three-fourths of the squirrels from the hard-cone region had a bony ridge atop their heads. This ridge helps provide a strong bite, since it gives the muscles that move the jaws additional anchorage. In the softer-cone region, less than half the squirrel skulls had the ridge. We might predict that coevolution in this community will produce ever tougher pine cones—and ever stronger squirrels.

Domesticated plants, or cultivars, are generally unable to grow husks or shells as thick as their wild relatives. Witness the difference between the cultivated English walnut and the American black walnut. Anyone with a good nutcracker can shell the cultivar, but a sledgehammer or mattock would be about the right equipment for shelling nuts of the wild black walnut. Euell Gibbons, in *Stalking the Wild Asparagus*, suggests driving your car over a bag of them!

Corn is an odd exception to the rule of less huskiness in cultivars. The wild ancestors of corn probably had husks, but they were loose enough to let the seeds out after ripening. In contrast, the corn we know now has its whole ear tightly wrapped in husks. The chance of a sprouting seed being able to force its way out is very small. The ears must be husked and the seeds sown by human hands. The underlying strategy for this kink in corn evolution is that the husks help protect the seeds from wild animals, saving them for us.

Self-defense through what might be called the paid bodyguard syndrome has been documented from a variety of plants. All of them manage to keep a horde of ants on themselves. These ants bite or sting any attackers, thus protecting their host from damage.

The first published description of the bodyguard syndrome appeared in the nineteenth century. Thomas Belt, an English engineer and enthusiastic naturalist working in Central America, noticed that the bull's-horn acacia tree harbored ants. This tree, common in the New World tropics, has fat hollow thorns near the bases of its feathery leaves. Ants live inside the thorns and in hollow twigs on the tree. At the tips of the leaflets, the acacias grow tiny pellets of nutritious material, now named Beltian bodies. The ants greedily devour the Beltian bodies. They eat no other part of the tree, though, and they attack any caterpillar or grazing mammal that tries to feed on a bull's-horn acacia. Belt commented, "I think that these facts show that the ants are really kept by the *Acacia* as a standing army."

Later biologists were not so sure. It was suggested in the 1930s that the ants exploit the trees, sheltering in the hollow thorns but giving no real protection. One biologist quipped that the trees got "as much use from their ants as a dog from its fleas." Conclusive evidence on the question came in 1969 with experiments in the field. Insecticide sprays were used to deprive several acacias of their ant hordes. Other acacias were merely sprayed with water, and kept their ants. The results were spectacular. The deprived acacias were far more heavily damaged by herbivores. And, in an unexpected side-effect, they were also overrun by neighboring vines and trees. So the ants not only protect their acacias from animal attack, but may even help them out in competition with plant rivals.

At least one kind of ant acts as a parasite of this mutualism. The tropical ant *Pseudomyrmex nigropilosa* can take over abandoned acacias, and although it likes to eat the Beltian bodies it fails to give its host any protection. But this feather-bedder should not be allowed to blacken the good names of the many ants that faithfully defend their trees.

Worldwide studies on the acacia family reveal a positive correlation between herbivore pressure and the possession of an ant bodyguard. Australia had very few large grazing mammals until Western contact, and none of its many species of acacia have Beltian bodies or keep ants. Most of them even lack the hollow thorns. There are plenty of ants in Australia, too, but none of them are involved in guarding these acacias. In the New World and the acacias do harbor ants, which can be seen as a response in Africa, where grazing mammals are common, most of to the pressures of their environments.

Other trees besides acacias attract and keep ant body-guards. *Cecropia*, a small tropical tree, produces its own little tidbits for ants to eat, known as Müllerian bodies. When the little Müllerian bodies ripen they are rich in glycogen, a nutritious compound normally found in animal tissues and highly attractive to ants.

Cecropia trees grow in South America, Central America, and on the Caribbean islands. Müllerian bodies and ant guards are both present in *Cecropia* in South and Central America and in some of the West Indies, but on the small islands between Tobago and Puerto Rico many of these trees lack such specialties for self-defense. This suggests that the original homeland of *Cecropia* trees is South America, where intense pressure from grazing mammals and insects promoted the evolution of the ant bodyguard syndrome. When *Cecropia* trees later spread north to islands with less herbivore pressure, they ceased producing Müllerian bodies and lost their superfluous ant guards.

Macaranga, another tropical tree that attracts an ant guard, produces yet another type of edible tidbit, now termed the Beccarian body in honor of the botanist Odo Beccari. The ant guards on *Macaranga* not only bite caterpillars and other herbivores on their trees, but also actively remove butterfly eggs laid on the trees, thereby helping to prevent caterpillar infestation.

Trees are not the only plant group to benefit from ant guards. The myrmecophytes, or ant plants, which often grow on trees, also support a standing ant army. Myrmecophytes belong to several different plant families, but they all look similar in one regard: each has a swollen, warty structure that houses ants. The part that forms the ant-shelter may be a greatly enlarged stem, root, or leafstalk. Ants living in these shelters benefit the plants substantially. Even when grown under ideal (herbivore-free) conditions in a greenhouse, these plants grow better if they keep their ants with them.

When ants that lived in myrmecophytes drank a honey-and-water mixture spiked with radioactive minerals, the minerals were soon found in the warty skin of the ant shelter. Later, the minerals spread throughout the plants. The ants, it seems, somehow help to nourish their host. In a similar experiment, ants living in a myrmecophyte were also offered fruit-fly larvae containing radioactive carbon. The carbon from these meaty tidbits was first digested by the ants and found later spread throughout the tissues of the host plant.

For many myrmecophytes, life is spent attached to a tree limb or on the side of a tree trunk. Since they are not true parasites, and cannot extract nourishment from the tree they cling to, the ant plants rely on rainwater and their ants for nourishment. In such a harsh microenvironment the tiny quantities of nutrients brought up by the ants can mean the difference between life or death for these plants.

Other plants attract and keep an ant guard without offering any shelter or solid food. Their come-on is nectar that flows from little glands scattered along the edges of leaves, on leaf-stalks, and on flower stalks—the extra-floral nectar glands. Like the Beltian, Müllerian, and Beccarian bodies, these shining droplets of sugary sap keep the ants coming back to the plant.

The tropical tree *Bixa*, which yields a yellow dye used

in coloring butter, attracts ants by such extrafloral nectaries; so do various species of wild morning glory throughout the tropics. Field observations suggest that the nectaries and the associated ants give these trees and vines substantial protection against herbivores. A field study that used the application of a sugar-and-water "artificial nectar" to plants that normally lack it has helped to demonstrate the value of an ant guard in the tropics. Several groups of beans were planted, and the leaves of half the plants in each group were sprinkled daily with sugar-water; the other beans were sprinkled with plain water. In clearings and along forest edges, the bean plants with artificial nectar showed much less insect damage than those sprinkled with plain water.

The prickly pear cactus in the American Southwest is another plant that attracts an ant guard by means of extrafloral nectar. A prickly pear guarded in this way can escape damage from sucking bugs that inflict extensive injury on ant-free cacti. In one study, the prickly pear nectar gland secretions were chemically analyzed and found rich not just in sugars but also in the ten amino acids essential for ant growth. Prickly pear is giving its bodyguard broth as well as wine.

Wild sunflowers have at least one species, and perhaps more, that can attract an ant guard with extrafloral nectar. In this case the glands are found on the large overlapping green scales, or bracts, under the flower. Ants feeding on the nectar there drive away flies that try to lay eggs in the flowers. Since the fly larvae hatched from the eggs would eat the sunflower seeds, the irritable ants are protecting the future of their plant. Here too the nectar supplies a rich mixture of amino acids as well as sugar. Indirect evidence of the value of an ant bodyguard emerges from the observation that insect damage to the sunflowers increased the further away they grew from an ant nest.

Ants, of course, are not unmitigated benefactors to their plants. A European study on birch trees underscores this point. Ants often place aphids, or plant lice, on birch trees, and lick up the sweet sap that the aphids extract from the trees. Indeed, the ants tend and protect the aphids, which do the actual dirty work on the trees. On the other hand, when populations of the leaf-eating caterpillar *Oporinia* became dense, the least amount of caterpillar damage occurs on ant-harboring birches. Thus a delicate balance exists between the ants' injury to the birches by pasturing their herds of aphids on them, and the ants' aid to the birches by driving caterpillars off of them. In years of heavy caterpillar infestation, though, the ants do come out as heroes.

One more beautiful example of a tree that employs ant guards is the wild black cherry found throughout New England. When the leaves on this tree unfurl in spring they sport tiny nectar glands along their edges and at the base of each leaf stalk. Later, nectar production in the leaf-edge glands tapers off but the stalk glands become more active. At the same time, ants are also becoming active after a long winter underground, and the eastern tent caterpillars are hatching. These voracious larvae love black cherry leaves and can defoliate branches or even a whole tree, if given a chance. But when the black cherry grows close to an ant nest, the ants are attracted to the nectaries; they attack any caterpillars that they encounter on the tree. The closer a black cherry tree is to an ant nest, the better it will stand up to caterpillar infestation. In this instance, the young caterpillars can be thought of as living tidbits of flesh that add to the ants' basic nectar reward for patrolling the tree.

CHAPTER 3

Self-Defense: Fleeing from the Wrath to Come

O generation of vipers!
Who taught you to flee
from the wrath to come?
MATTHEW 3:7

INSTEAD OF ARMOR, BARBED WIRE, OR BODYGUARDS, certain plants have evolved a deceptively simple defense strategy. They run away from their enemies.

It seems odd at first to think of a plant doing this. Nearly every biology text ever written announces in its first chapter that animals are more mobile than plants. (The examples given, of course, are usually an eagle versus an oak tree, not a barnacle versus a seaweed.) For most of us, it takes the time-lapse photography of Disney's *The Living Desert* to convince us that plants can move at all. If they are so slow, how can they flee from their enemies?

The answer lies in the life history of these evasive plants.

They are the weedy plants that spring up in disturbed soil, flowering one day and fading the next. They are the plant world's version of mayflies, short-lived and small. The American ecologist Robert MacArthur coined the expression "r-selected" for the pattern their lives display: rapid growth, early maturity, a high rate of reproduction, and a short life. Quick to grow and quick to reproduce, they often do outrun their enemies by being here and gone before they attract any notice. Good examples would be ragweed and fireweed, familiar plants in newly turned or newly burned soil. Among garden plants, most of the familiar annual flowers would also fit this pattern, such as nasturtium, petunia, and marigold.

The r-selected plants are able to flourish in a wide range of soils and rainfall. They are reminiscent of the pioneers who settled our western frontier. Tough and adaptable, they know how to make the best of a bad situation. But like some of the historical pioneers, they are less at ease in settled surroundings, and this may be because they are not specialists. They are thus at a disadvantage when forced to compete with specialized plants. In a long-established, stable biological community such as a mature forest, the dominant plants are usually not the r-selected species. Instead, large-size, slow-growing, long-lived plants dominate—trees, shrubs, woody vines, and flowers such as goldenrod and aster. MacArthur dubbed this group the K-selected species.

Because K-selected species are slow to mature, they are initially at a disadvantage in competing with r-selected weeds. Their offspring take several years to appear and even then are scanty at first, as anyone who has planted and impatiently watched a fruit tree can testify. But when

K-selected plants do mature and start to set seed, their seeds are almost always larger, and each parent plant lives so long it has many batches of offspring instead of the single batch that an r-selected plant leaves behind.

In certain communities, the r-selected species may have a brief moment of glory. California poppies blazing on desert soil after a rain, or the dandelions on a solid-gold lawn, are familiar examples. Their abundance shows that short-lived, small, and weak though they are, they manage to remain one of the most successful plant groups on earth.

The animal groups that pose the biggest threat to r-selected plants are those that have evolved a similar life history. These animals follow the annual weeds in hot pursuit through their growing season. Caterpillars and aphids are among the best-known of such short-lived, small, rapid-growing creatures. A female aphid can produce thousands of aphids in a few weeks if she lands on a suitable plant, by the process of parthenogenesis (a method of self-duplication in which all you need is Mother). A caterpillar, though it cannot multiply its own numbers, can double its body weight several times in a few weeks. Both can devastate the soft-stemmed succulent new growth of their food plants.

Behavior has evolved to sophisticated levels in these little herbivores, allowing them to shadow their short-lived hosts. When three species of caterpillar that feed primarily on r-selected plants were investigated in the laboratory, each proved to have a pattern of searching for food that neatly matched the natural distribution pattern of its own preferred host plant. They may not always get their plant but they are certainly hot on its trail.

Certain plants practice a more covert lifestyle than the

r-selected plants, one that helps conceal them even more from the unwelcome attention of their enemies. They spend a large portion of the year underground, generally reduced to a bulb or fleshy tuber. Then they make a brief, explosive and beautiful appearance above ground during the flowering season, as for example tulips, daffodils, and crocuses. Few animals attack the flowers or leaves of these short-lived beauties. Many species of fungi follow a similar life pattern, with a long-lived underground net of tissue called a mycelium, from which at rare intervals a large, bright, above-ground mass—the mushroom—rises.

The seeds or spores of r-selected plants are small and light. They can be carried great distances by the wind. Like the seed broadcast by the sower in the New Testament parable, some strike dry stony ground, others are choked among thorns, and only a few fall in good places. But, as if to compensate for this hit-or-miss method of seed dispersal, most r-selected plants produce huge numbers of seeds. Since the wind carries them far from their origin, they may at least escape the notice of animal enemies that have located their parents.

In addition, many r-selected plants produce seeds that can go dormant, or hold off on sprouting, if the conditions where they land are temporarily inappropriate. A classic study begun in 1934 by burying ragweed seed in bottles of sand revealed that even forty years later some of the seeds could still germinate when given the right combination of season, temperature, and moisture.

Another ecological stratagem for escaping animal antagonists is the specialized life-history known as mast-cropping. Mast, an Anglo-Saxon word, applies to nuts that ripen and fall in such huge quantities that swine can stuff themselves on the crop. Often the crop appears at unpredictable intervals. Beech trees in the Old World are classic examples of masting; their nuts are still called beech-mast. Piñon pine and Douglas fir in the western United States, and hickory trees in the East, are good New World examples of mast-cropping. But the most spectacular case is seen in Oriental bamboo. One species of bamboo common throughout Asia goes for about 120 years between crops. A Chinese expression for a very elderly person is, "He has lived long enough to see the bamboo flower twice."

The value of mast-cropping, according to the American ecologist Daniel Janzen, is its tendency to satiate all possible predators on the seeds. More seeds are released at once than can be devoured, and thus some will survive to sprout the next year. A parallel from the animal kingdom is the seventeen-year cicada; underground and invisible while immature, the adults emerge synchronously in such huge numbers that all their enemies are soon satiated. Thus some of the newly emerged cicadas can survive and reproduce the species.

When bamboo flowers and sets seed in the wild, it may be on such a massive scale that acres of forest floor are blanketed with the seeds. Huge numbers of wild jungle fowl and wild pigs come to gorge themselves on the unfamiliar feast. Janzen suggested that crowding together at these nutritive hot spots preadapted wild ancestors of domestic poultry and swine to being tamed. Their insensitivity to crowded eating conditions makes them easy to house in captivity, and their acceptance of easily stored grains makes it simple to provide their food.

Another animal associated with bamboo is the giant

panda of southern China. This distant relative of the raccoon feeds almost entirely on the young sprouts of bamboo. In the year after a mast crop of seed is formed and shed, the bamboo forest dies. No new sprouts appear until a few years have passed. The normal panda food source is wiped out, and catastrophic drops in their population follow. The panda also has to contend nowadays with human population pressures that have reduced its natural forest environment. Like a braider of carriage whips after the advent of automobiles, the panda may be a specialist whose day is over, fated to survive only as a conversation piece in zoos.

A few trees are now known to practice a third ecological stratagem for enemy evasion: they drop off leaves and fruit injured by animal attack. This technique, in which the sacrifice of a part helps protect the whole tree from destruction, was first documented in holly trees. Leaves infiltrated by leaf-mining caterpillars are dropped from the tree, and the infestation fails to spread. Since the holly lacks a nervous system, the interesting question is how the trees can tell the difference between their healthy and infested leaves.

A similar response to insect damage is exhibited by hickory trees, which jettison injured nuts rather than leaves. In a field study of hickories that showed a heavy and premature summer dropping of nuts, the vast majority of the nuts remaining on the trees turned out to be healthy. Those which had been dropped revealed a heavy level of infestation with insects. It looks as if these trees are acting to reduce any further investment in their injured offspring. Here again in the hickories, as in the hollies, the tantalizing question is: How does a tree differentiate between healthy and injured tissues?

A few researchers of animal–plant antagonisms have focused on the question alluded to earlier in regard to ants and aphids: can herbivores both damage and benefit their hosts? In some cases, the answer to this question may be yes; research on aphids provides an example. As mentioned earlier, an aphid drains sugary sap from its host plant, thus depriving it of energy-rich material. Aphids usually suck far more sap than they can digest, and as a result they constantly drop sugary sap, or aphid honeydew. Anyone who has parked a car beneath an aphid-infested tree in early summer has probably cursed their copious rain of sticky fluid. (When dry, the honeydew forms white sugary flakes. Certain Biblical scholars believe that the manna mentioned in the Old Testament was in fact dried honeydew—it was "white, and sweet as sugar.")

But the honeydew is especially rich in one particular sugar known as melezitose. When melezitose enters the soil, it promotes the growth of nitrogen-fixing bacteria that add to soil fertility. Thus it may well be, at least in this case, that the plant receives injury and aid from the same animal.

ΡΕΤΣΙΝΑ
ΜΕΣΟΓΕΙΩ
ΑΤΤΙΚΗΣ

Self-Defense: Repellents, Poisons, and Contraceptives

FROM POISON IVY TO THE STRYCHNINE VINE, THERE IS a wide array of plants that secrete irritating, distasteful, or downright poisonous chemicals. They practice chemical warfare against their animal antagonists, warfare in which nerve-damaging, cancer-causing, and fertility-destroying drugs have all been documented. But it took us a long time to see plant poisons in this light. For many years they were considered useless, irrelevant compounds—biologically meaningless, except that now and again one would turn out to have medical value. A recent writer on poisonous plants crystallized this attitude when he remarked of digitalis, a bitter secretion by foxglove, "The poison has no known purpose [for the plant]." In light of the hard chemical work the plant must do to secrete such a compound, the statement is about as hard to swallow as foxglove tea. When a creature can repel, sicken, or kill its antagonists, it is at a distinct advantage in the struggle for survival. Natural selection as a result has favored chemically armed plants; they are abundant in all dry-land communities.

It must be admitted that even biologists, for years, left the defensive chemicals of plants ill-studied and misunderstood. Because these compounds are such a mixed bag in their chemical structure and in their effects on the eater, they were labeled "secondary compounds," and were considered of minor importance in plant life as compared with sugars, fats, or proteins. Then, in the late 1950s, the zoologist Gottfried Fraenkel hammered out the theory we accept now as obvious truth: these so-called secondary substances defend plants against animal attack.

How did such substances originate? Even now, many aquatic plants lack defensive chemicals, common though they are in land plants. The very earliest plants, of course, were all aquatic. When some came up onto dry land, they were exposed to bright sunlight untempered by a watery umbrella. Even though sunlight drives the photosynthetic process essential for green plants' survival, too much sunlight is dangerous—it can bleach and kill green leaves, as any houseplant tender knows. Some newly emerged land plants accidentally evolved the ability to secrete colored compounds that acted as blockout agents against the intense sunlight. These compounds, termed flavonoids, are still widely distributed in land plants, and it turns out that they are unpalatable to animals. So plants that secreted flavonoids reaped a dual reward: screening from sunlight and some protection from animal attack.

Later, as green plants flourished and multiplied on land, plant-eating insects and reptiles multiplied too. Under the pressure of such voracious herbivores, those plants that had the genetic good luck to come up with more powerful repellents were bound to succeed better. The earliest of these additional weapons were the compounds known as tannins. Bitter and puckery to the taste, tannins make a plant unpleasant to eat—and hard to digest once eaten. They tend to inactivate animal digestive juices. The few modern plant survivors of the dinosaur era—bracken fern, tree ferns, cycad palms—are loaded with tannins. So are quite a few more modern plants, among them oak trees and the Oriental tea plant.

Some plant-eaters seem to have iron-clad digestive systems. Pigs, squirrels, and certain woodpeckers happily devour raw acorns, which are rich in tannins. But the human cultures that used acorns for food, like the North American Indians, treated them to wash out the tannins. The simplest technique, according to one modern forager for wild food, is to shell the nuts and let them sit in cold running water overnight.

Few if any modern cultures use acorns for food, but almost all do prepare tea to drink. Tea is often promoted as an innocuous substitute for coffee. Indeed, tea is touted as a beverage that anyone—even an invalid—can digest. But a recent study suggests that the tea plant is capable of a quiet revenge on its major antagonist, the human species. The botanist Julia Morton discovered that cancer of the esophagus is most common in exactly those cultures that drink black tea most heavily. Especially striking is the case of a northern province of China where strong black tea is mixed with rice to form a gruel eaten daily. Esophageal cancer rates reach frightening levels in that region, and so far no medical writers have advanced a compelling explanation of the cause. It seems likely that lifelong ingestion of the tannin-loaded gruel is at the root of this esophageal cancer epidemic.

A manipulation of tannin-rich tea practiced mainly by the British is the addition of milk or cream to each cup. The protein in milk combines with the tannin and makes it less potent and harmful to the drinker's gut.

Other beverages rich in tannins are dry red wines, cider made from high-tannin apples, and certain herbal teas. These beverages, too, are associated with a high rate of cancer of the esophagus. Even beer—usually made from a mash of low-tannin grain—can become high in tannins when aged over beechwood chips. This is actually a common practice in the brewing industry, since the increase in bitterness provided by the wood chips gives the beverage an astringent quality many beer drinkers like. In an interesting social sidelight, most of the regions where beer or wine is the major source of tannin report a rate of esophageal cancer anywhere from two to ten times higher in men than in women. We can safely interpret this sex-related difference as a reflection of the fact that in most cultures the men drink larger quantities of beer and wine than the women—though this may be changing nowadays.

Even sorghum, a grain widely raised for cattle fodder in this country, can secrete high levels of tannin, and this in turn affects the health of the cattle that eat it. High-tannin strains of sorghum are favored by many farmers because they are so resistant to insect and bird devastation. But when high-tannin sorghums are used for fodder, the cattle raised on them show on average a twenty-percent reduction in body weight, compared with those raised on a low-tannin strain.

Another type of animal-repellent, digestion-damaging secretion comes from the group of resinous plant compounds known as terpenoids. These are the substances that give a pine forest its sweet, pungent smell, and retsina wine its peculiar taste. Unlike tannins, which are present in low concentrations in young leaves but in high concentrations in older leaves, terpenoids are usually abundant throughout the life of a pine needle. But the relative concentration does vary from tree to tree. In a field study of ponderosa pines, researchers noted that tassel-eared squirrels ate the needles only from certain trees in each grove, leaving others untouched. When twigs from the exploited and the untouched trees were analyzed, the untouched trees turned out to be far higher in terpenoids. The squirrels were simply behaving as well-adapted herbivores should: pursuing the pine of least resistance.

Another pest on ponderosa pine, the tiny insect known as black pine-leaf scale, shows a similar ability to discriminate between high- and low-terpenoid trees. A field study of the problem found the needles on some trees coated with the scale, while neighboring trees were completely free of it. When the researchers tried to transfer the scale insects from branch to branch, they succeeded if the move was to another branch on an already infested pine. If the attempted transfer was to a branch on a non-infested pine, it usually failed. The little scale insects have only adapted to specific trees—presumably those lowest in terpenoid content—and many generations of their short lives pass before the tree's death makes invasion of a new one necessary.

The resinous, sweet-smelling wood of red cedar has long been valued in our culture as a repellent to the invasion of clothes moths. Now an even more effective insect-repelling wood has been reported from a leafy tree: the neem tree, native to India. Traditional wisdom there dictates hanging neem leaves over exposed food, and placing neem-wood beads near stored grain. It turns out that the leaves, seeds, bark, and wood of neem are all rich in a compound known as azadirachtin (from the Persian name for neem tree). Azadirachtin is highly repellent, even poisonous to many insects, but is not harmful to vertebrate animals. Neem extract may have commercial value as a highly specific insect-repellent compound.

Tannins and terpenoids are mainly secreted by trees. The soft-stemmed or herbaceous flowering plants have generally evolved a different system of chemical defense, shared with a few fungi: they secrete toxins. These compounds sicken or kill the unwary herbivore, instead of merely repelling it or disturbing its digestion. Some toxins are bitter, but others give no advance warning of their presence by any bitterness or peculiarity in taste. The classic example of this most dangerous situation is the deadly species of *Amanita* mushrooms. *Amanitas* are frequently mistaken for edible mushrooms by careless or over-confident foragers. An American book on edible and poisonous mushrooms published in 1895 painted an unforgettable picture of three agonizing deaths in one family whose members had eaten a species of *Amanita.* Surviving relatives were questioned; they had eaten smaller quantities of the mushroom stew that killed the victims. They described it as quite tasty.

Flowering plants and their toxins became abundant toward the end of the dinosaur era. At least one biologist in fact has suggested that large-scale poisoning by toxic plants new to their surroundings was the smoking pistol that extinguished the dinosaurs—an imaginative but untestable idea.

Why do large plants—trees and shrubs—generally secrete digestibility-reducing compounds like tannins, while the shorter-lived, smaller plants secrete toxins? Part of the answer is that toxins are simpler in chemical structure—they cost the plant less to make. In addition, they are effective at lower doses than tannins and terpenoids. Thus they are cheaper for plant gypsies to tinker up in their short encampment on earth.

Large, long-lived species such as trees can afford the investment necessary to secrete high levels of the more complex tannins and terpenoids. And if a tree secretes a high enough level of these compounds, it is protected from a wide range of enemies. A toxin, however, may provide effective defense against only a few herbivores. Toxins can almost always be deactivated if the herbivore involved can secrete the right digestive enzymes.

As several biologists have pointed out, this means that only short-lived plants, "unapparent" to their enemies, can afford to use toxins. Trees and other long-lived plants, in contrast, are around long enough to be quite apparent; they are "bound to be found." Their best strategy is broad-spectrum chemical defense, as with terpenoids and tannins. Huge doses of these compounds may accumulate in a tree: tannins can make up as much as sixty percent of the dry weight of an oak leaf. This pattern of difference in chemical defenses parallels the life-history distinction between r-selected and K-selected species: the r-selected plants go in for the lightweight toxins, the K-selected species for the heavier weaponry.

As with tannins, certain toxins can be carcinogenic, and this is true of the toxins in one of the most innocuous-tasting of all cooked vegetables—the parsnip. Whether raw or cooked, parsnips are rich in compounds known as psoralens. These can cause gene mutations and promote the occurrence of cancer in laboratory animals, and in humans they can cause a peculiar sensitivity of the skin to sunlight—followed by a higher chance of developing skin cancer. Oddly enough, psoralens combined with ultra-violet light form an effective treatment for psoriasis, a troublesome skin disease, and this approach to psoriasis is currently a popular one in some circles. The possible exchange of psoriasis for skin cancer seems like a poor bargain to the medical conservative.

Some toxins can be removed from plant tissue by heat, so that cooking thus renders the plant safe to eat. Though this is not true of psoralens in parsnips nor for the neuro-toxins in *Amanita*, it is true for a wide range of toxins in shell beans and soybeans. Soybeans are an especially dramatic case in point. Raw, they are loaded with compounds that can interfere with human digestion and cause dangerous clotting of the blood. Cooked, they are so harmless and rich in digestible proteins that soybean products—soymilk, curds, and tofu—are popular foods throughout Asia. Some toxic wild greens such as milkweed can also be rendered harmless through cookery. Often rich in a group of potent heart poisons known as cardiac glycosides, milkweed greens in the kitchen should be subjected to three successive baths with boiling water. After this torture-like treatment, the cooked greens are mild in flavor and appear to be low in cardiac glycosides.

Recent evidence shows that toxins, like tannins and terpenoids, also differ in concentration from one plant to another. Milkweeds vary a great deal in this regard. So do clovers, some of which release hydrogen cyanide when injured by a herbivore. The clover plants that can release this poisonous gas are termed cyanogenic, while the ones that fail to do so are known as acyanogenic. When clover is only lightly grazed, the acyanogenic forms do as well as the cyanogenic ones. But when the plants are stressed by heavy grazing, whether by cattle or slugs, the unarmed clovers suffer more damage.

Slugs are among the most abundant grazers on clover. As we might expect, the acyanogenic forms are few where slugs are numerous. But one study sheds an interesting sidelight on the problem. In response to such stresses as extreme cold, transplantation, or exposure to disease-causing fungi, the acyanogenic clovers survive better than their poisonous relatives. These results are a useful reminder that animal–plant coevolution is far from being the only factor that influences plant survival.

Many plants show a neat relationship between their chemical defenses and the presence of animal antagonists: the plant secretes the defensive chemicals only after it is attacked. We might compare this situation to the military policy of a nation that maintains no standing army, but has a draft law enabling it to create one rapidly once attacked. Cyanogenic clovers, in a sense, are doing this when they wait to release cyanide gas until after being bitten; but in this case, their troops have already been marshalled. In a number of trees, the defensive secretions are simply not produced at all—or else are secreted only in tiny quantities—until circumstances warrant upping the level.

Birch trees studied in northern Finland follow this pattern. A young birch tree may be so low in chemical defenses that it can be overrun and defoliated by caterpillars. But if it survives, its levels of chemical defense become so high for the next three years that few caterpillars can live on it.

Further north, snowshoes hares in the Arctic are able to devour twigs and the main shoots of young birch and alder. But after the main shoot is gnawed off, these little trees send up a group of new shoots, or suckers, from the roots. The suckers are loaded with bitter chemicals that stop the hares from wreaking any further damage on the new growth. In fact, a watery extract of the ground-up new growth when sprayed onto palatable twigs repels the hares.

Even the familiar oak tree and its tannins employ a stratagem of this sort. The year after an oak has been severely damaged by gypsy moth larvae, it grows new leaves that are far higher in tannins and in fiber than the foliage on uninjured trees—almost as if to say "You won't get away with this twice." The tremendous crashes in gypsy moth populations that follow such repulsive peaks as the one in New England in 1981 may be the natural result of caterpillars starving to death on the trees that their grandparents' attacks turned unpalatable.

Willow trees have recently been implicated in a particularly effective strategy of chemical defense. In one study, the willows damaged by caterpillars showed an elevated level of repellent chemicals in their new leaves—no surprise in light of the previous studies. But so did even undamaged willows that grew nearby! The current hypothesis is that some airborne chemical wafts a warning message from the injured trees to their neighbors, readying them to resist attack. For a Tolkien reader, this process is reminiscent of Old Man Willow angrily calling up other trees to threaten the little band of hobbits within his territory.

Many plants show a seasonal fluctuation in levels of chemical defenses. Oaks are normally low in tannins in the spring and higher by late summer. In field and laboratory study of winter-moth caterpillars, which feed on oak, birch, and willow, the larvae strongly preferred young leaves over old ones, especially favoring the early spring growth. And when batches of the caterpillers were raised in the laboratory, the groups reared on spring foliage showed substantially better growth.

Leaf-cutter ants, a major pest in the tropics, chew up huge masses of leaves into green confetti, and carry the pieces to their deep underground dens. But they are discriminatory in their devastation: they avoid the bitter leaves of coffee trees, and they always prefer to take the younger leaves of the plants they damage. Since young leaves, low in defensive chemicals, are also softer and moister than the old leaves, the question is still open as to whether the ants choose the leaves on the basis of taste, texture, or water content. These ants do not eat the leaf confetti itself, but use it as compost for their real food, a tiny fungus; this partnerhip is discussed in more depth in Chapter 9. The leaves may be gathered in accordance with what suits the fungus and not the ants.

Soil composition in each community, according to one theory, should also play a part in determining levels of chemical defenses in the plants that grow there. The assumption is that plants growing on poor soil will secrete larger quantities of defensive chemicals, because under-nourished plants are faced with a harder time repairing or surviving animal damages. Thus natural selection would especially favor plants under these circumstances that are able better to repel attacks. Field study of two African rain forests supports this belief. One forest grew on acidic, nutrient-poor soil, the other on less acidic, richer soil. Leaves from trees in the first forest were loaded with phenolic compounds, powerful deterrents to herbivores. Leaves from trees in the second forest had very low levels of these secretions.

The study also examined feeding behavior of *Colobus* monkeys in both forests. *Colobus* are technically leaf-eating monkeys, but the group in the high-phenol forest ate fewer leaves than those in the low-phenol forest, favoring seeds instead, which are less repellent-loaded. They were also very choosy about the leaves they did eat. They avoided many of the common trees and vines in favor of rarer, lower-phenol species.

Another study, this one in India, focused on feeding behavior in the langur, another leaf-eating monkey. The results clearly showed an avoidance of tannin-rich leaves. The langurs fed almost exclusively on two or three species of low-tannin trees. Surprisingly, though, they ate several species of plants rich in alkaloid compounds usually considered poisonous to mammals. They survive this apparently lethal diet, the researchers believe, because of bacteria they harbor in their guts, which detoxify certain poisons.

A much larger mammalian leaf-eater, the North American moose, also harbors such symbiotic bacteria in its gut, but in this case the bacteria are shielded from poisons by the animal's careful selection of items in its diet. A study of the feeding behavior of free-ranging moose on Isle Royale in Michigan revealed that they generally chose plants richer in nutrients than the average plant in the area. Yet they almost never ate the leaves from certain species—even though the avoided trees were also high in nutrients. The plants they shunned included several very rich in repellent chemicals, and at least one of these compounds, methyl salicylate, is harmful to the symbiotic bacteria of the moose. Since these symbiotic bacteria digest the tough fiber that makes up a substantial portion of the moose's leafy diet, they are essential for its survival. So its dietary finickiness is an adaptive pattern for the moose.

Strikingly different in action from either tannins or toxins are the plant secretions that act like animal hormones. The best-known of these mimics—the insect hormone analogues—may cast what seems like an evil spell on an insect's life. Understanding these potions calls for a brief plunge into the physiology of insect growth. Many insects, such as grasshoppers, grow up gradually, changing through a series of molts from tiny juvenile to bigger, essentially similar-looking adult. Others, such as moths, undergo a dramatic metamorphosis from juvenile to pupa to adult. In both cases, the juveniles are mobile and voracious, but only the adult is fertile and sexually active.

It was shown in the 1950s that a juvenile insect will

stay juvenile and sterile as long as certain cells in its head keep on secreting a hormone appropriately called juvenile hormone. Secretion of this hormone normally declines with the passage of time, and others replace it. A drop in juvenile hormone allows maturation to begin.

Many plants, the best-known being balsam fir, are now known to secrete compounds that mimic insect juvenile hormone, and thus prevent insects from developing into the sexually mature stage. This forever-young effect was discovered by accident when a group of bugs studied in the laboratory were placed on cage bedding made of newspaper. They all failed to mature, even though they had already turned off their own production of juvenile hormone. It turned out that the newspaper, manufactured from balsam fir chips, was loaded with what is now termed a juvenile hormone analogue. As long as the bugs were in contact with it, the bedding kept them juvenile—and sterile. Such analogues function to reduce future populations of an insect antagonist.

Some biologists were enthusiastic about juvenile hormone analogues as ideal insecticides to replace the broad-spectrum organophosphate insecticides still widely used. The latter can wreak harm on vertebrate animals as well as on insect pests. But there are two problems with the analogues: they are costly to purify from plant sources in the large quantities needed for commercial application, and they also allow the juvenile insects to go right on eating as long as they last. Although the affected insects will never reproduce, they can devour a lot of plant material before they run out of steam.

A more promising technique may come from recently discovered plant compounds that interfere with metamorphosis in the opposite direction: they bring a pre-mature end to the juvenile stage. The plant compounds involved are known as "precocenes" because they force precocious maturation on affected insects. Precocenes were first identified from the small flowering plant blue ager-atum in the 1970s; they have since been documented in a number of other species. The premature "maturity" that they bring about makes for a miserably small, sterile adult with malformed mouthparts, which soon dies. The pre-cocenes do this by chemically clinging to, and interfering with, the glandular cells in the insect head concerned with normal production of juvenile hormones.

Insect herbivores are not the only victims of hormonal counter-attack—sterility has also been forced on warm-blooded animals by hormone mimics in their food plants. As long ago as the 1940s, an Australian study showed that a wildflower often eaten by sheep made the ewes sterile. The culprit was a clover secretion very similar to the vertebrate sex hormone, estrogen. Although estrogen helps both initiate and maintain female sexuality in all birds and mammals, overly large quantities disrupt ovula-tion. And when the sheep devoured the clover containing large quantities of the estrogen mimic, now termed phyto-estrogen, they failed to ovulate.

More recently, plant-induced sterility has been docu-mented in the United States in a small wild mammal, the mountain vole. Voles, sometimes called short-tailed mice, are common small rodents of open fields and meadows. The voles studied live in mountain meadows in the west-ern U.S., where they feed on grasses. The young plants in spring are free of phytoestrogens, but by late summer and early fall they are loaded with them. The heavy dosage of hormone is unaffected by travel through the voles' gut. It enters the bloodstream and soon shuts down ovarian activity—a plant-initiated birth-control program. Vole populations plummet as a result.

If we look at the overall process from the voles' point of view, we could say that these results are in their own best interest. Reproduction races ahead in the spring, when vegetation flavor and quantity are at their peak. In fact, researchers found in a more recent experiment that young, growing plants are rich in a chemical that actually stimulates female ovulation and male testicular growth in the voles. Thus vole population increase is rapid in spring, as their food plants give the chemical green light. But it slows to a crawl in the fall, when hard times lie ahead for small plant-eating mammals.

From the plants' point of view, their ability to impose temporary sterility on the voles is a valuable weapon in the struggle for survival. Late summer is when many of the plants form their seeds, tempting tidbits of protein, starch, and oil to a hungry rodent. By forcing a crash in the vole population at just about the same time, the grasses protect their precious offspring—the seeds—from total destruction.

A study of sagebrush and California quail revealed a similar interaction between plant and herbivore: control of reproduction in the bird by phytoestrogens in the brush. The sagebrush, in fact, rules quail fertility with an iron rod. In years when rainfall is high, the plants are free of estrogen mimics, or low in them. But in years that follow a drought the plants become rich in phytoestrogens—and the quail populations drop drastically. Here again the plants are showing a finely tuned ability to hold down herbivore numbers: after a year of good rainfall the sage-brush can eaily replace parts nipped off by quail, but after a year of drought they grow poorly and benefit greatly from reduced quail populations.

Even as commonplace a plant as wheat, especially winter wheat, is capable of secreting phytoestrogens. These are usually found in the germ rather than in the starchy part of a wheat kernel; what long-range effect these hormone mimics have on the human consumers of the wheat crop is as yet unknown.

Human beings have managed to exploit certain plant defensive secretions very ingeniously. Steroids from a species of yam serve as raw material for synthesis of oral contraceptives in pharmaceutical laboratories. Strychnine, atropine, and curare, all potent plant toxins that affect animal nervous systems, have been used as chemical tools in neurobiology research. Digitalis from foxglove has long been a familiar medicine to treat congestive heart disease.

In a grimmer vein, such poisonous plant secretions as curare have been used in hunting and in warfare, usually by dipping arrow or blowgun darts into a gummy extract of the plants. And at least in fiction, plant toxins are favorite murder weapons, as in Arthur Conan Doyle's *The Sign of the Four* and Mary Webb's *Precious Bane.*

In most domesticated plants or cultivars, the capacity for chemical defense has been weakened or wiped out entirely. Not in all, however. Potatoes still secrete substantial quantities of neurotoxins in their above-ground stems and leaves; raw cashews are loaded with a powerful skin irritant; and raw beans can cause a dangerous tendency to blood clotting. But it is more often true that, as with the physical defenses considered earlier, chemical defenses are considerably reduced compared to wild relatives. Cultivated cabbage, for example, has much lower levels of repellents than its wild relatives in the mustard family. As a result, cultivated cabbage is far more palatable to us than the wild mustards are, but it is also more palatable to insect pests that compete with us for the food plant. Domesticated plants are a captive, nearly disarmed, population. Modern farmers confront plant pests on a huge scale largely because our ancestors succeeded so well in breeding tasty—and vulnerable—cultivars.

CHAPTER 5

Self-Defense:
One Man's Meat
Is Another Man's Poison

As with physical defenses and life-history strat-egies, chemical defenses are never effective against all antagonists. It is not too surprising that some animals can handle a plant poison that strikes down others. But what is really surprising is that plant chemicals that can repel or kill some animals actually attract others. The defensive chemicals secreted by plants in the mustard family—cauliflower, Brussels sprouts, kohlrabi, kale, cabbage, as well as mustard itself—are a good example. The chemical weapons of such plants are known as mustard oil glycosides, and these glycosides certainly repel many plant-eaters. If young aphids that normally feed on bean plants are blown by the wind onto mustard family plants, they begin to feed where they land by sticking their sharp little beaks into their new host. But if they find even tiny quantities of mustard oil glycosides in the sap, they pull out their beaks, open their furled wings, and sail onward.

One group of researchers forced swallowtail-butterfly caterpillars to eat food rich in mustard oil glycosides. They found that the compounds can be downright unhealthy for the creatures that normally avoid them. Swallowtail caterpillars usually select food plants in the umbellifer family—wild carrot, parsley, Queen Anne's lace, and celery. In this study, they were raised on celery stalks. Each stalk had its base resting either in plain water or in water mixed with sinigrin, a compound that releases mustard oil glycosides. The celery gradually drew up the fluid in its vessels to the point where the caterpillars fed.

The swallowtail caterpillars that fed on celery resting in plain water grew into normal adults. Those raised on celery in low concentrations (0.001 to 0.01 percent) of sinigrin grew into adults that were less fertile than normal. But those raised on higher concentrations (0.1 to 2.5 percent) of sinigrin never matured at all—they died as caterpillars.

And yet sinigrin, so toxic to swallowtail caterpillars, is practically the staff of life to the larvae of cabbage butterflies. There are several species of these little creatures. Long ago it was shown that adult females of a cabbage-feeding species strongly prefer to lay their eggs on plants rich in mustard oil glycosides. The larvae, when hatched, feed exclusively on these leaves and mature beautifully.

More recently, researchers on other cabbage-feeding species of butterfly found that caterpillars of one species, *Pieris brassicae*, chew on leaves if and only if their oral sense organs detect mustard oil compounds. In another species, *Plutella maculipennis*, the caterpillars will try to eat up pieces of filter paper soaked in a solution of mustard oil glycosides. In both these species the larvae thrive on cabbage plants, maturing into fertile adults.

Even human beings occasionally show a perverse preference for the pungent flavor of mustard oil glycosides. These compounds are what give prepared mustard its snap; mustard is one of those foods that seem to be more highly valued the hotter they are on the tongue. And in traditional Western medicine, the same compounds give mustard plasters their skin-blistering quality, supposed to be valuable in clearing up throat and chest disorders. In a more exotic context, folk dancers in southern India put mustard seeds under their eyelids before performing. The physical and chemical irritation from the seeds reddens their eyes and gives them a (literally) inflamed look, suggestive of intense emotion.

The legume family, which peas and beans belong to, also includes a number of species poisonous to certain animals. In a study that used the tiny cowpea weevil as animal subjects, black beans were examined for their toxicity. The beans were ground into a powder, then mixed with the cowpeas on which this weevil normally feeds. Black beans are rich in a toxic compound known as phytohemagglutinin (PHA for short), which promotes a lethal clotting of the blood. The cowpea weevils whose food was contaminated by a very low concentration (0.1 percent) of the black bean PHA grew fairly well, though their survival rate was slightly lower than normal. Other batches of weevils fed on cowpeas with from 2.5 to 5 percent PHA all died before maturing. Yet cowpeas contain their own array of toxins, none of which deter cowpea weevils from doing well on them.

This study might raise some disquieting thoughts about the safety of black bean soup. Luckily for us, it turns out that heat inactivates the PHA, making bean soup harmless. One ecologist has even suggested that human cookery originated in the need to detoxify abundant and nutritious seeds of wild legumes.

The evidence is convincing that certain animals have penetrated the chemical defenses of certain plants, but why should any actually prefer a plant loaded with repellents or toxins? One possible answer is that a small animal that feeds on a plant avoided by others will escape the risk of being inadvertently swallowed by a large herbivore. This is a definite survival benefit! In addition, of course, a small creature thereby avoids competition for the same food with larger herbivores.

The idea of escape from competition also suggests the concept of exclusive feeding rights. Once an animal evolves the inheritable ability to detoxify a poisonous plant, it makes sense for it and its descendants to specialize on that plant. The presence of the poison ensures that the plant is still harmful to other creatures and thus remains an exclusive food supply.

Among caterpillars, the task of deactivating plant poisons is usually performed by digestive juices known as oxidases and epoxidases, sometimes lumped together under the initials MFO (mixed-function oxidases). And in their feeding behavior, most caterpillars can be placed in one of two large categories: polyphagous or eaters-of-many, and oligophagous or eaters-of-few. The caterpillars of painted lady butterflies are a good example of the first category: they devour thistle, sunflower, alfalfa, and hollyhock as well as other plants. Caterpillars of checkerspot butterflies illustrate the oligophagous life-style: they dine only on the turtlehead plant and its close relatives. Caterpillars of the familiar monarch butterfly are another oligophagous group: they eat only milkweeds and dogbane.

Polyphagous caterpillars ingest a wider variety of plant toxins than oligophagous ones, yet they grow just as well. Does this mean that they compensate for their toxin-rich diet by producing extra quantities of detoxifying MFOs? An extensive analysis of MFO secretion rates in many species of caterpillars indicates that the answer is yes. Significantly higher rates of MFO output were detected in the polyphagous species than in the oligophagous ones.

A further question of interest is whether a low MFO output could be made to increase. One group of researchers hypothesized that this could be done by feeding caterpillars in the low-secretion group a diet richer in varied plant toxins than they would normally encounter. They found that, indeed, this exposure to unaccustomed plant poisons turned the trick in several species. Levels of MFO secretion rose markedly. The outcome might be compared with the changes seen in human beings who repeatedly consume yeast toxin (ethyl alcohol to the chemist). Exposure to alcohol induces higher levels of the enzyme alcohol dehydrogenase by the liver, so that the toxin is digested more rapidly in the experienced drinker. The novice drinker may get a hangover from one glass, while a practiced toper scarcely notices any effect at all.

An intriguing side effect of chemical warfare in co-evolution is that many animals are not only able to devour plants poisonous to others, but they are also able to store up the plant toxins in their bodies. Like the weapons of a defeated foe, these secretions can be useful to the animal as a defense against creatures that threaten its own existence.

In some cases the captured weapons are put to use very simply and directly. Pine sawfly larvae, a major pest of needle-bearing trees, are able to accomplish rapid recycling of the terpenoids from their food plant. After chewing the needles, they reserve most of the terpenoids in a special pouch at the back of the mouth. (How they decide what to swallow and what to store is still unknown.) If they are disturbed by a potential enemy, they spit out this pungent fluid and thereby repel most of their would-be predators.

Caterpillars of the painted lady also spit green fluid when disturbed. Although this has not yet been analyzed, it seems likely to contain compounds that are also predator repellents, derived from the bitter juices of their food plants.

Most people can remember catching grasshoppers as children and getting their hands stained with a brown froth the captives give off, known in New England as grasshopper molasses. The desert locust, a large grasshopper that becomes a major agricultural pest when it emigrates from crowded origins, produces the same brown froth when disturbed—and it is loaded with terpenoids. Derived from resinous desert plants on which the locusts feed, the pungent froth repels most insect-eaters. (Anyone feeling sorry for John the Baptist, living on a diet of locusts and wild honey in the desert, will be pleased to know that according to modern Biblical authorities what he really ate was carob pods and wild honey.)

A beautiful example of biological opportunism has been documented in a study of the defensive froth of another species of grasshopper. In this case, the froth was loaded with man-made herbicides known as dioxins or 2,4,5-T and 2,4-D. These compounds, used as jungle defoliants in Vietnam under the name of Agent Orange, have also been widely used as weed-killers in the United States. These grasshoppers apparently ate vegetation sprayed with these herbicides, survived, and stored them to use against their foes—turning human sprays into insect weapons.

Undoubtedly the best-studied case of recycled plant poisons is seen in the monarch butterfly. This big black-and-orange long-distance traveler, known from Nova Scotia to California, is talented at chemical weaponry as well as extensive flight. Fragile though they look, every fall monarchs fly down the West Coast from Vancouver to southern California, and down the East Coast from Canada to central Mexico, to spend the winter in warm quarters. The Canadian biologist Frederick Urquhart, in his book *The Monarch Butterfly*, details the long years of butterfly-banding and geographical detective work that went into his discovery of the East Coast wintering ground. A still-concealed location in Mexico, the trees decked with monarchs there are described as "an enchanted forest" by the American biologist Lincoln Brower. Brower and his associate William Calvert are among the very few people other than Urquhart ever to visit and photograph the Mexican wintering ground of the monarchs.

One might imagine the two biologists chatting beside the same campfire late at night while the wind rustled through the wings of monarchs in the trees, but in fact this idyllic picture would be quite mistaken. Urquhart regarded Brower and his associates as interlopers, and criticized their field techniques. Brower's award-winning film on the monarch, *Strategy for Survival*, achieves a gentle revenge: Urquhart's name is not even mentioned once.

More important from the point of view of animal–plant coevolution is the chemical recycling practiced by the monarch. While a caterpillar, it devours leaves of milkweed and dogbane plants. These two plant families are rich in bitter toxins known as cardiac glycosides. Akin to the digitalis in foxglove, cardiac glycosides act on the heart of warm-blooded animals and cause greatly increased contraction, which can lead to death from heart failure. A lucky feature for birds or mammals that consume such plants by accident is that cardiac glycosides are also nauseating, so that the animals may vomit the poison before they are killed by it.

Yet monarch caterpillars feed on these toxic plants with impunity. They store the toxins, retain them through metamorphosis, and still possess them as adult butterflies. A careful analysis reveals that by far the heaviest concentration of toxin is in the orange-and-black wings—precisely the parts most likely to be nipped at first by a would-be predator.

This acquired poison protects the butterflies against at least one kind of predator. Brower offered monarchs to wild-caught bluejays in his laboratory; the jays pecked at the monarchs' wings but then released them, beak-marked but alive. (Occasionally you can see such a beak-marked butterfly in the wild. It looks as if someone has snipped a little wedge out of its wings—the scar is mute evidence of bird attack.) In a few cases, the jays pulled off the butterfly's wings and ate the body. Although much lower in toxins, the body was still poisonous enough to nauseate the jays. After eating they vomited, and after vomiting they generally refused to try monarchs again.

The combination of conspicuous coloration and poisonous or unpalatable flesh, such as the monarch shows, results in conspicuous, warning actions, known as aposematic—literally "off-signed"—behavior. Aposematic animals move slowly, deliberately, with the colorful individuals often congregating so that they are even more conspicuous. Aposematic animals are defended by their distastefulness, and this is advertised—to the experienced predator—by their colorful conspicuousness. The slow-moving, brilliant hues can be translated as "I taste bad—stay away."

Other butterflies are aposematic like the monarch, thanks to their toxic food plants. Swallowtail caterpillars feed on wild carrot, celery, parsley, and Queen Anne's lace, as mentioned earlier, and nearly all of the wild umbellifers are rich in a plant poison known as furanocoumarin. (The poison Socrates drank came from water-hemlock, yet another plant in this poisonous family. Hemlock extract seems to have acted as a nerve poison in his case.)

Furanocoumarin affects plant-eating insects not adapted to umbellifer food plants in a different way: it makes them vulnerable to sunlight. The furanocoumarin weakens chemical bonds in the genetic material of life, DNA. Then, if exposure to sunlight occurs, the light disrupts DNA structure; illness or death follows. As one group of researchers points out, this gives us a new perspective on the habit certain leaf-eating insects have of rolling themselves up in their leaves. If they are eating umbellifers, as long as they stay out of the sun they escape the deadly effect of the poisons.

Despite the danger wild umbellifers pose to insects, swallowtail caterpillars actively select these plants for food. In fact, they will choose a plant rich in furanocoumarin over one low in it whenever they have an option. They obviously gain almost exclusive feeding rights thereby. Do they gain another benefit as well? Adult tiger swallowtails are large yellow-and-black, highly conspicuous butterflies. Are they aposematic like the monarch? Yes, according to a laboratory study carried out in Europe. When swallowtail caterpillars were offered to wild-caught insect-eating birds, the birds pecked at the caterpillars but ate none of them. Most of these birds even refused to peck at them a second time. The question of whether furanocoumarins are retained through metamorphosis and concentrated in wings of the swallowtail butterfly awaits a reply in a comparable laboratory investigation of adult swallowtails. The fact that one occasionally sees a beak-marked swallowtail in the wild suggests that, like monarchs, they do keep their larval poisons.

We should acknowledge here that some aposematic animals, such as the skunk, manufacture their own horrible-smelling defenses from scratch. And at least one brightly colored, distasteful insect group, the chrysomelid beetles, manufacture their own repellents regardless of the foodplant they are reared on. If a monarch or swallowtail is deprived of its poisonous food plants, does it lose its distastefulness, or can it, too, manufacture its own toxins from scratch?

Dietary manipulations carried out on the monarch show that the source of poisonous protection for it is indeed the foodplant alone. When monarch caterpillars were reared in the laboratory on cabbage—a difficult task since some of them starved rather than eat it—they turned out to be acceptable meals for captive bluejays. A more natural parallel of this experiment has been documented as well. Certain tropical milkweed vines actually secrete no toxins at all. Monarchs raised on these undefended vines are just as defenseless themselves when they reach adulthood.

To turn the question around, we can also ask if it is possible to make a harmless animal into a poisonous one, or at least an unpalatable one, by changing its diet. This has been accomplished in the laboratory with tobacco hornworms, the fat jade-green caterpillars of a common sphinx moth. The hornworms normally eat tobacco, tomato, and related plants, all pungent and bitter but not toxic enough to make the hornworms poisonous.

Three batches of tobacco hornworms were reared in the study. One group was fed on a semisynthetic formula rich in powdered milk and nutritional yeast—a sort of "Let's Have Healthy Hornworms" diet—which included no food plant material. The second group was fed on tobacco, the normal food plant. The third group was fed on belladonna, a plant related to tobacco but rich in atropine, a potent plant poison.

All three groups thrived and entered the chrysalis or

pupa stage. Just before pupating, the caterpillar usually voids most of the chewed-up plant material from its gut. The tough, shiny hornworm pupae were offered to hens, which gobbled them up. The hens which ate pupae from the first and second groups of subjects showed no signs of illness, but the hens which ate pupae from the third group all died. The hornworms were able to store up atropine in their flesh. If the experiment were to be repeated, it would be interesting to raise all three groups to maturity, to see if the poison is retained in the adult moth—and, if so, in what body parts it is concentrated.

Many biologists have assumed that aposematic behavior calls for the sacrifice of a few individuals in every generation, so that predators can learn to avoid the species. Bitten, pecked at, or gulped down whole by a hasty predator, the individual victims would suffer for their conspicuousness. But survivors in the same species would benefit since they would be avoided later on by the sadder but wiser predator.

This concept of aposematic defense fits in well with a currently popular theory in biology—the concept of kin selection. This theory, first outlined in 1964 by the British biologist William D. Hamilton, attempts to resolve a paradox in evolution. Darwin's idea of natural selection seems to demand that each individual struggle to improve its own chances of survival and reproduction. Yet there are many examples of animal altruism, behavior that imperils the individual while it benefits the group to which the animal belongs. Darwin's contemporary, the Russian writer Peter Kropotkin, gathered many examples of such behavior in his book *Mutual Aid.* One of the best-known is the raising of an alarm call in birds. The bird that spies a predator approaching, and gives a call that alarms the flock, thereby raises its own risk of death—but the flock as a whole is benefited. Aposematic behavior seems like another example. Colorful bodies and slow movement draw attention to an individual butterfly, which can hardly benefit if it is bolted down by a predator.

Kin selection theory explains such behavior on the basis of a genetic payoff to the victim's family. If altruistic behavior is inheritable, then an altruist's closest relatives are likely to share its genes for self-sacrifice. When one member of such a family risks its own neck but thereby raises the chance of survival for its relatives, the whole group—the kin—benefit. Such a family or kin group is at an advantage in the long struggle for survival over nonaltruistic families. The family whose members direct altruistic behavior toward one another is likely to be favored and to multiply. In a semifacetious vein, one biologist summed up the theory of kin selection with the words "I will lay down my life for four brothers or eight cousins." The more relatives you have surviving you, the more copies of your genes survive, and thus any aspects of your behavior that are genetically controlled will also survive and multiply.

A major aspect of this theory criticized by biologists unsympathetic to the idea of kin selection is the assumption that altruist genes will out-compete genes for selfish individualism. Wouldn't individuals with the latter makeup out-compete the altruists? Kin selection adherents deny that this is so. They point out that the self-sacrificing behavior is, in their theory, always directed toward relatives, so it benefits one family lineage—not the whole society.

Such a lineage could soon outnumber those whose motto was "Every man for himself and the devil take the hindmost."

Another problem with kin selection theory, however, is its strong emphasis on degrees of relatedness. With many animals in the wild, it is very hard for the human observer to be sure of family relationships. For that reason alone there are many biologists who reject the idea of kin selection. They see it as being based on assumptions of relatedness that are difficult or impossible to prove.

One last difficulty with kin selection theory as an explanation for aposematic behavior is that, at least at first glance, the concept seems to demand that the sacrificed butterfly intend to help its relatives. Even if it becomes possible to show—and this has yet to be done—that the death of an individual monarch in a bluejay's beak does lead to its relatives escaping future attacks, how can we believe that the victim wanted this to happen?

Proponents of kin selection reply that of course the butterfly didn't intend to be caught, but its genes dictated that it raise its chance of being caught—by its colorful, conspicuous flight—since copies of the same genes in relatives will benefit.

But a simpler solution than kin selection theory may account for aposematic behaivor. As mentioned earlier, bluejays that sampled monarchs tend to peck at their wings and then frequently release them. Though beak-marked, these butterflies can survive to reproduce, and the beak-mark itself might keep experienced jays from trying such a butterfly again. Thus the individual itself would benefit. And in the study on swallowtail caterpillars, we saw that insect-eating birds pecked at the subjects but did not eat them. All of the caterpillars survived, apparently unharmed. Aposematic animals, in short, may survive predatory attack better than we used to believe. Their colorful distastefulness may be explained, after all, by benefit to the individual.

Some animals that are not aposematic have evolved defenses that depend for their success on an evolved resemblance to plants; this strategy is known as cryptic behavior. A form of mimicry, it can include, as in a walking-stick insect, slow swaying motion, color, and shape—all suggestive of plant rather than animal life.

A form of animal defense that involves resemblance to a poisonous animal is the strategy known as Batesian mimicry. Here, a harmless animal has evolved a resemblance not to a plant but to an aposematic animal. Widespread in the tropics, Batesian mimicry is best-known in the northern temperate zone from the interaction between monarch and viceroy butterflies. Viceroys are not related to monarchs; they eat different food plants and are in fact quite palatable to birds. But an adult viceroy looks quite like an adult monarch. Both species have big orange and black wings. As long as the two species live in the same community, birds that taste and reject monarchs will also avoid viceroys.

A theoretical requirement for the success of this system, it has been said, is that the model species—the unpalatable or poisonous one—ought to be more numerous than the mimic. Only then will it be more likely that an average predator will taste a bitter victim and decide to avoid any creatures that resemble it. Field studies have found that monarch and viceroy populations actually meet this theoretical requirement. Where they coexist, the monarchs are more abundant than the viceroys.

It must be acknowledged that even aposematic behavior, like the plant defenses on which it often rests, is not effective against all enemies. Laboratory studies on monarchs several years ago showed that mice, quail, and several other animals could devour even the butterflies' poison-laden wings with impunity. More recently, Mexican farmers who pasture their cattle near the East Coast monarch wintering site have smoked butterflies down in huge numbers to feed to their cattle as a sort of high-protein dietary supplement. The cattle, far from keeling over, show no ill effects. Several birds that live in the same vicinity, including orioles, feed on monarchs as well. Is this because they have evolved the ability to detoxify cardiac glycosides, or do they select victims low in poisonous protection? Were the bluejays used in the original laboratory studies abnormally sensitive (for birds) to the poisons in their victims? As yet these questions remain unanswered.

Biologists have looked in the direction of the learning process to explain how animals come to discriminate against unpalatable foods. The British biologist W. H. Thorpe suggested that learning be defined as "an adaptive change in behavior based on experiences." There is no denying that refusal to recapture a bitter or poisonous creature is adaptive—that is, beneficial—to any animal. But psychologists have been dissatisfied with this explanation, since the two types of learning usually studied in the laboratory—operant conditioning and classical conditioning—both fail to account for animal rejection of foods that induce illness or nausea. In both types of conditional learning, it is proverbial that a punishment or negative reinforcer should follow hard on the heels of the behavior to be eradicated. In psychological lingo, there must be contiguity of reinforcement. And this seems to be true even for higher animals: any experienced dog-trainer can attest that breaking a dog of a bad habit can only succeed when the punishment comes within seconds of the action. Yet many poisonous plants and animals affect their consumer minutes, hours, or even a day or two later. How can wild animals learn to avoid these delayed-effect items?

Research with laboratory rats by the psychologist John Garcia has led to an answer. His original work was provoked by a practical side of rat behavior little-studied until then by psychologists—the question of how rats escape pest-control programs. Such antirat programs have been conducted ever since human beings settled down, started raising grain, and stored it in bins. Grain storage in ancient Egypt was plagued by rats, and so is grain storage in modern Kansas. Apart from such time-honored interventions as sending in ferrets or fox-terriers, or setting out traps, the most popular approach has long been the use of poisoned baits.

Rats will eat almost any poisoned bait initially. It may kill quite a few of them (and if the farmer is lucky, they die and rot outside of the grain bins). But after the first few nights the bait is taken less often. Eventually the survivors totally reject the poisoned bait, and the rat population begins to rebuild. The poison, whatever it was, is just as toxic as ever, but the rats will not eat it any more.

Garcia decided to investigate this bait-shyness, as

rodent-control people call it, in a laboratory setting. He raised his rats on a diet of standard lab chow, then offered them small amounts of a new food. As they consumed the food, they were subjected to a heavy dose of X-radiation. The new food was actually harmless, but the radiation made the rats become severely ill a few hours later. When they were offered the same new food a few days afterward, they refused it.

Garcia suggested that this adaptive change in behavior is a type of conditional learning that he named aversive conditioning or food-aversion learning. It is now often called the "Garcia effect" in his honor. Natural selection, in his theory, has forced wild animals to be acutely sensitive to the effects of the foods they eat. They remember, and later avoid, the last new food encountered just before the sensation of illness began. Such learning would be especially relevant for rats, which consume a wide variety of foodstuffs—and for human beings, also traditionally omnivorous. Human experience suggests that we, too, go through aversive conditioning: we remember with revulsion the food that preceded a bout of illness.

In later research on food-aversion learning, rats were reared in social groups in the laboratory rather than rearing each rat in its own cage. Only the older rats of the group underwent aversive conditioning; but *all* members of the group learned to reject the novel, illness-inducing food. Young rats that had never even seen the food until offered it in testing, rejected it just as firmly as the older rats that had experienced illness after eating it. The logical assumption is that social life allows information about food selection to be passed from the older to the younger generation. It is tempting to imagine a gathering something like the Rescue Aid Society in Walt Disney's *The Rescuers*, where an elderly rat leans back on a pile of cedar shavings and holds forth to an audience of respectful young rats. "I tell you it was horrible. Just escaped by the hair of my head, I did!" But a simpler process of imitation may be all we really need to explain the results here. Young rats may simply eat what they see their elders eat, and avoid anything they do not see used.

A case of aversive conditioning has been described in the lowly slug. Small but important as a grazer on many pasture plants, the slug shows behavioral responses similar to those Garcia obtained with rats in his first experiment. The slugs in this study were offered more than twenty different foodplants in order to find which they preferred. As any animal behaviorist working on a tight budget would have predicted, they liked cultivated mushrooms, the most expensive food offered, by far the best. They were then deprived of food long enough to get really hungry and later offered bits of mushroom to eat. Just after they started gnawing on their pieces of mushroom, the researcher gassed them with moist carbon dioxide. This made the slugs sick. Afterward, they rejected mushroom almost to a slug.

From the viewpoint of aversive conditioning, the ideal chemical defense is one with poisonous qualities and a distinctive, easily learned flavor or smell. There would be no need, in this light, for a plant or animal to actually be deadly fare for its enemies. Nausea or severe indigestion would work just as well to educate would-be consumers. Many plants secrete compounds that conform to this ideal

—but not all of them, as we saw with furanocoumarins and with *Amanita* mushrooms. We might consider those plants guilty of using an elephant gun to shoot a mosquito.

Lilies of the Field: Flowers and Cross-Pollination

Consider the lilies of the field,
how they grow; they toil not,
neither do they spin, yet I tell you
even Solomon in all his glory
was not attired as one of these.

LUKE 12:27

WHEN WE TURN TO THE PEACEFUL SIDE OF COEVOLU-
tion with the examination of animal–plant mutualisms,
the best-known example is certainly pollination. Pollination syndromes almost always benefit plant and animal
partner alike (except for certain cases, to be discussed later,
in which the plant has evolved a little fast dealing with
its animal partner).

Pollination leads to the setting of seed and is the major
route to reproduction in land plants—but not the only
one. Many land plants such as iris, cattails, and bamboo,
spread by underground stems that throw up new shoots.
A single plant can blanket an entire field with offspring
created through this simple self-cloning. Or in other cases,

a branch broken off a woody plant may fall into moist soil, send out roots, and grow. In fact, nineteenth-century accounts of fence-building along rivers in the Midwest sometimes mention a cottonwood or willow fencepost growing into a tree. In still other cases, the parent plant may even set adrift tiny versions of itself grown at its leaf margins, like the desert plant called "mother of thousands."

Aside from such asexual techniques, usually termed vegetative reproduction (although it is not restrained to vegetables—sponges, corals, and segmented worms practice it, too), there are also some green plants that manage to pollinate their own flowers. They can thus set seed in isolation. Such selfing, as plant breeders call it, can occur in tomatoes, peas, and many others.

But in the long run this isolationism and vegetative reproduction are likely to have a bad outcome. Charles Darwin said "Nature abhors a perpetual self-fertilization," and the statement still appears to hold true. Cloning and self-fertilization represent extreme inbreeding. They keep genetic diversity low in the population. If every environment were completely stable, of course there would be no harm in this. But geologists tell us that all environments undergo constant change. Ponds fill in, hillsides erode, rivers change their courses, pastureland becomes desert. In a world of flux, a genetically static plant is not favored to endure.

The genetic diversity that allows a species to evolve in a changing world is most easily attained by outbreeding. In outbreeding, two separate individuals each donate a set of genes to their common offspring, a set of genes that is a new assortment of the parents' original set. Outbreeding, in human experience, demands a temporary union of two separate individuals. This, of course, is true of other land animals, too. But many marine animals and all wind-pollinated plants use a more casual technique. Their sex cells are released into the environment, to drift together and unite on their own. The parents never touch each other, though they do have to synchronize their breeding cycles to ensure simultaneous release of sex cells.

Flowering plants that depend on animal aid in pollination achieve sexual reproduction through what might be called artificial insemination by donor. The pollen grains are collected from one flower and transferred to a spot where they can unite with the female cells, the ovules, in another flower—ideally, on another plant of the same species.

It has long been assumed that for flowering plants the casual, hit-or-miss process of outbreeding by wind pollination preceded the more structured, sophisticated process of animal-aided pollination. This time-honored assumption, however, may be wrong. Two of the best-known researchers into floral biology, Karl Faegri and Leendert van der Pijl, believe that animal-aided pollination or zoophily (animal love) is just as ancient as wind pollination or anemophily (wind love). Zoophily may, in fact, be the older technique. Among modern flowering plants, the cycad trees most closely resemble the plants of the dinosaur era, and they are insect-pollinated.

At about the same time that flowering plants began to flourish on land, insects began to multiply, too. Coevolved mutualisms between the two groups can account for much of the success each has experienced since that early day. Many of our modern wind-pollinated plants, especially the grasses and coniferous trees, bear signs of insect pollination in their ancestry. These plants are commonest in dry or cold environments. Pollination ecologists have suggested that as land plants spread north from the moist tropics which abound in insect life, a few plant families came up with adaptations for wind pollination. These lucky few would have been the ones to thrive in the chilly North—as spruce, fir, and pines now do.

A problem that many plants—wind or insect-pollinated —must face occurs when the same plant produces both male pollen and female ovules. This creates a risk of self-fertilization. Such hermaphroditic structure is overcome in many plants by the timing of sexual development. In certain plants the pollen grains ripen long before the ovules do, a condition called protandry or male-first. In others, the ovules ripen long before the pollen, a condition known as protogyny or female-first. Both strategies reduce the risk of self-fertilization.

Pollen-producing plants are grouped by botanists into two categories, the gymnosperms (naked seeds) and the angiosperms (covered seeds). The angiosperms all have flowers with protective and decorative structures around the ovules. Although the flowers may be tiny in wind-pollinated species, angiosperm ovules are always hidden by covers known as carpels. Gymnosperms, in contrast, lack carpels; their ovules are exposed to the air.

Carpels of angiosperms usually fuse into one central chamber, the pistil, topped by a smokestack-like tube, the style. At the top of the pistil is a receptive surface, the stigma. Pollen grains that arrive at an angiosperm flower cling to the stigma and send a tube down through the style. In many flowers, the pollen grains from a different plant of the same species sprout their tubes faster than pollen from flowers on the same plant—another neat stratagem that helps guard against self-fertilization.

The pollen grain slips two cell nuclei down through its tube into the ovary. One nucleus fertilizes an ovule, and this gives rise to an embryo. The other pollen nucleus fertilizes a female nucleus near the ovule; this grows into endosperm tissue that nourishes the embryo.

In addition to these basic structures, most angiosperms also create a circle of petals or corolla, ringing the female pistil and the male pollen-producing stamens. Many have a ring of modified leaves, or calyx, outside the corolla. Both layers help to protect the reproductive organs. More important from the coevolutionary viewpoint, in animal-dependent flowers the corolla, and occasionally the calyx as well, draw the attention of pollinators with brilliant colors and eye-catching shapes.

Angiosperms are by far the most abundant plants on land today, except in cold northern regions where gymnosperms like spruce and fir take over. Many theories have been proposed to explain the numerical dominance of the angiosperms. Their closed carpels have earned great credit, as they provide protection against insect damage to the ovules. Since the earliest insect visitors to land plants were probably beetles, which chew and gnaw indiscriminately, ovule protection would have been a distinct advantage.

But closed carpels have another, more subtle effect. They ensure that, of all the pollen that lands on the plant, only a few grains will get to the ovules. The pollen grains that most vigorously extend their tubes down through the pistil will win in the race for the ovules. As we already saw, this competition favors pollen from other individuals of the same species over self-produced pollen. In addition, when pollen of a separate species hits the stigma it is totally inhibited, or slowed down, in the growth of its tubes when compared with grains of the same species. Thus the flower escapes not only inbreeding from self-pollination, but also escapes sterile interspecific hybridizing, through the protective action of the carpels. The flower wastes little time, thanks to the carpels, on creating inbred or sterile offspring.

A recent reconsideration of the virtues of closed carpels has underscored the point that, under this arrangement, pollen grains compete with each other in the race to the ovules. The competition will be more severe, of course, the more pollen grains are available. If genes in a pollen grain can express themselves in the vigor with which the grain extends its tube, then genetically superior pollen grains will presumably win this race. So closed carpels may help promote intense natural selection for the cream of the crop among pollen grains. In gymnosperms, the absence of carpels means that the naked ovules must accept the first pollen grains that hit them. Unless there is some chemical incompatibility, the first grain of pollen to land on an ovule fertilizes it, whether good, bad, or indifferent. The resultant offspring might well lack in genetic vitality, thus coming in second to the angiosperms in competition on dry land.

The hypothesis that rivalry among pollen grains results in better offspring was tested on petunias. Three groups of the flowering plants were pollinated by hand. The first group received a scanty amount of petunia pollen on their stigmas, the second group a moderately generous amount, and the third group a superabundant amount. Competition among pollen grains was at a minimum in the first group, since barely enough pollen was supplied to fertilize the ovules. In the third group, competition was at a maximum. Offspring of the third group of petunias, when raised to maturity, grew significantly faster and larger than the offspring of the other two groups. Increased pollen grain competition does seem to lead to better offspring.

A closed carpel, like a green jewel case that allows a flower to conceal its treasured ovules, may have started angiosperms along the road to ecological success. But petals and sepals, almost equally ancient floral structures, benefit angiosperm flowers in a different way: they attract animal attention. Except in wind-pollinated angiosperms, which have reduced corollas and calyces, the petals and often the sepals are conspicuously colorful. Most botanists agree that petals and sepals originated from green leaves that grew near the carpels. But petals now come in a dazzling array of colors, which allows them to stand out against the background of leaves.

Insects, which have color vision, began their own history of ecological success at about the same time as the angiosperms. It takes no great stretch of the imagination to picture the first few colorful flowers attracting the enthusiastic attention of primitive insects, and reaping a rich reward of cross-pollination. Birds and mammals, later animal groups with color vision, have also made significant contributions as pollinators.

But color is not the only lure coevolved by flowers under the pressure supplied by animal pollinators. Fragrance can be equally important, sometimes even more important. The light, sweet fragrances of fruit-tree blossoms, roses, and other flowering plants are attractive not only to us but to bees. Bees avoid putrid odors, but other insects do not. Flies love rotting meat; it provides just the right food for their offspring to live on, and they avidly search out its smell. In an elegant dovetailing of coevolution, certain flowers that fail to attract bees are able to seduce flies into pollination visits by offering them the sort of putrid stench they love.

Food, in liquid or solid form, is a bribe that many flowers have evolved to secure pollinator services. The common reward is the sugary sap we call nectar. Although nectar is associated in most people's minds only with floral biology, it seems to have been originally produced by glands scattered over primitive ancestral plants. Even today, some plants like bracken fern release nectar although they have no flowers. And many flowering plants, as we saw in Chapter 3, release nectar from extrafloral glands. But the floral nectar secretion is generally much greater.

Streaks, spots, and patches of contrasting color within a flower announce the location of the floral nectar glands, or less commonly the location of pollen; they offer a visual marker for the pollinator. Not by accident, these guide-marks are placed so that the would-be drinker of nectar must brush against the sex organs of the flower in passing. Many of the guidemarks are visible to human beings as well as to the pollinators. We can see the brown streaks on a purple rhododendron flower, the golden "eye" on a forget-me-not. But in some cases the marks are written with invisible ink as far as we are concerned. Marsh-marigold petals look uniformly bright yellow to us, but to the insect eye they each have a dark patch at the base. This creates a visual target directing insects to the center of the flower, the site of nectar, pollen, and ovule production.

Guidemarks invisible to us are the result of plant tissue that reflects ultraviolet light strongly. Our eyes, insensitive to ultraviolet as a color, fail to notice its reflectance, but insect eyes detect it. Oriental poppies, a brilliant scarlet to us, reflect ultraviolet light strongly and attract honeybees in this way. Otherwise, honeybees generally ignore red flowers.

Nectar in the flower, however abundantly it may ooze out, is the product of laborious chemical synthesis in the plant's leaves. After it is moved to the flower, it is usually protected there against accidental loss by evaporation or rain. Christian Sprengel, an eighteenth-century Lutheran pastor who was the pioneer investigator of pollination, was struck by the frequent presence in flowers of protective hairs, scales, or spurs that retained the nectar. He saw them as evidence of God's forethought and careful design. Without necessarily adopting his belief, we can see such structures as advantageous in the struggle for survival—a kind of safety-deposit box for the pollinator payroll.

Some flowering plants, instead of offering nectar as pay for the pollinator's services, give pollen; such plants produce huge quantities of pollen grains. This may have been the technique of the most primitive flowering plants. Wild roses typically produce such surfeits of pollen. Cultivated roses have been so subjected to human selection for more and better petals that their pollen production is very reduced.

Pollen as a food reward demands that the pollinator be able to chew, since each grain has a tough outer coating that guards its nutritious contents. Bees are among the few insects that utilize it, and even they reject pollen once adult. They chew it thoroughly before feeding it to their young, as human mothers in pre-blender cultures used to do with bread before feeding it to babies.

So-called bee pollen consists of solid little pellets of pollen grains stored in beehives when floral supply exceeds larval demand. Bee pollen has come into the limelight in recent years as a food supplement for human athletes. Some runners take it religiously, a sort of communion wafer that they hope will confer power, if not grace, upon them. But the tough outer coating of each pollen grain, the fact that human digestive enzymes have little or no effect on it, and the fact that the stored pollen is not chewed by adult bees before being stored, combine to make it likely that any benefits from such intake are entirely in the mind of the consumer. This, of course, may influence athletic performance in itself.

Flies, moths, and butterflies, with their thin tongues and weak jaws, usually cannot eat pollen. But a recent study of the large tropical *Heliconius* butterflies reveals that they gather pollen on their tongues, and dip it into nectar before swallowing it, like dunking a doughnut into a cup of coffee to soften it.

A few modern plants offer their pollinators what seems like a cruder, more primitive reward than either extra pollen or abundant nectar: they let the pollinators eat their petals. Magnolia trees do this, and so does sweetshrub, a fragrant flowering bush common in the southern United States. Both magnolia and sweetshrub are pollinated by beetles, which have large heavy jaws and a tendency to gnaw anything edible they can find on a plant. They find a rich reward in the edible petals, at least in sweetshrub. Biochemical analysis has shown that the inner petals of sweetshrub are loaded with protein-rich cells. These same protein-rich cells are also formed along the tips of the stamens, the pollen-producing structures. Beetles eager to consume these tidbits crawl into the center of the flower, inadvertently coating themselves with pollen grains, and then carry the pollen with them as they fly off to the next sweetshrub for another snack.

Beetles, most biologists assert, were the first insect pollinators. The present-day flowers pollinated by beetles have primitive forms: large, thick petals, and many of them. Their fragrance has a heaviness about it, apparently more attractive to beetles than the light sweet odors of fruit tree blossoms. A few beetle-pollinated flowers, in fact, smell rank rather than sweet—elderberry is a good example. The flowers have simple shapes, in which the beetles forage around carelessly. B. J. D. Meeuse, a vivid contemporary writer on pollination, compares beetles in flowers to bulls in a china shop—although bulls in an interior decorator's shop might be even more appropriate considering the silky texture of the petals.

Recent inspection of certain primitive flowering plants in New Guinea forests has led to the suggestion that flies, as well as beetles, were among the pioneers in pollination. These forest flowers are small, sweet, and release nectar, but their petals are peculiarly thick and fleshy. They attract both flies and beetles. And flies as well as beetles are frequent visitors to water lilies and tulip trees, both regarded as classical beetle flowers. Since flies as well as beetles were abundant at about the time that flowers began their evolutionary explosion, there is no compelling reason to accept beetles as the only pioneer pollinators.

A wide variety of modern plants are pollinated by flies.

In some cases, the flies appear to be substitute partners for plants that have moved north, away from their tropical bee pollinators. Most orchids are tropical and depend on bees, but one species found in northern bogs—ladies' tresses—is pollinated by mosquitoes that breed in the cold waters of the bogs. And a member of the milkweed family found growing wild in desert areas, where bees are rare, is another fly-dependent creature: the starfish flower, named for its big, five-petaled, horrible-smelling blooms. Starfish flowers play a deceptive game: they produce no nectar, no extra pollen, no protein-rich cell masses, but they do produce a rank odor that suggests rotting meat. This smell attracts female flies who take the flower—usually dark and hairy-looking as well as putrid—for a decaying carcass, the right place to lay their eggs. Their egg-laying visits transfer pollen from one starfish flower to another. Later, when the seeds have set, the hapless fly larvae can be seen wiggling hungrily in the rank-scented flower that exploited their mother.

Many fly-pollinated flowers combine a putrid odor with reddish-brown or greenish-brown color, like rotting flesh. A few are more sophisticated. Meeuse showed that the flower of the voodoo lily, a relative of our familiar calla lily, actually heats up when the pollen is ripe. This remarkable feat, due to a sudden speeding of cellular respiration in the deeper tissues of each flower, helps broadcast the odor of the flower far and wide, a putrid stench that attracts flies. In addition, the voodoo lily is lined with oily, slippery cells down which the careless fly slides, to be held captive deep in the flower until it is throughly dusted with pollen. After that, the slick cells, although still oily, go soft and

limp, and the fly can climb out by clinging to their surfaces. Slow to learn, the pollen-coated fly can easily be tempted into another voodoo lily, and thus effects cross-pollination.

Even more sophisticated fly-traps are built into the structure of other flowers. Dutchman's pipevine shows one of these: the light window. A thin, transparent area in the wall of a large, deep flower, the light window serves to attract captured flies to just the right part of their temporary dungeon for them to be dusted with pollen, then find their way out. Another lure, seen frequently on fly-pollinated orchids, is the growth of whiskers or fringes on the edge of the flower. Ladies' tresses have these, and so do several tropical orchids. In a breeze, the movement of the little whiskers on the flower suggests the presence of tiny flies on it. Though not highly social, flies do like to aggregate, and they tend to settle on flowers that appear to already have some visitors of their own kind.

A slightly different visual lure that also exploits the flies' sociable tendencies is seen in the tropical orchids known as Venus's slipper. These flowers are liberally dotted with small spots of black or brown. To a casual human observer, they appear to have flies all over their surface. A passing fly seems to get the same impression: it stops to visit a flower so obviously attractive to other flies. It ends up trapped in the slipper-shaped, oily-lined flower, to be released only after pollen transfer.

Could such exploitation have been the basis for an even crueler one, the digestion of flies by the leaves of insect-eating plants like Venus-flytraps? These plants grow in nitrogen-poor soils and use the digested protein of their victims' bodies as a self-supplied fertilizer to promote their own growth. Was there a time when their flowers were fly-pollinated? If so, the trapping of the flies might have extended to the leaves, but evolved into a truly devouring love.

An Israeli orchid has now been identified that, like Venus's slipper, has abundant small dark spots on its surface. But in this case the orchid attracts predatory insects known as hoverflies, whose larvae feed on aphids. The flies apparently see the spotted orchids as loaded with aphids. The female hoverflies lay their eggs on the orchids as well as on other plants that actually do have aphids on them. Here too, as with the deceived flies on the starfish flower, the misplaced larvae are doomed to starvation while the flower sets seed.

Flies have also been useful in the pollination of some cultivated plants. Onion flowers, for years hand-pollinated by human beings, are now less laboriously pollinated by blowflies. Cocoa trees, visted by a variety of insects, appear to depend for most of their pollination on midges. These little flies breed in the rotting leaves and fruit on the forest floor. An overly zealous groundskeeper who keeps the earth clean of debris deprives his trees of their pollinators.

A peculiar case of fly-pollination is found in the tropical orchid *Trichoceros*. These flowers, which grow at high elevations in the Andes, look very much like the females of a local species of fly. Males of this species visit the *Trichoceros* and attempt to mate with them. They eventually give up, but carry off a bundle of pollen as they fly away. Slow to learn, they try again when they see another *Trichoceros* flower, and thus effect cross-pollination. This futile intercourse has been dubbed pseudocopulation.

Butterflies—highly mobile and eager searchers for nectar—visit a wide range of flowering plants. Their presumed floral partners have sweet scents and often possess a long, narrow spur or tube that holds the nectar. The shape of the spur is designed to force the butterfly's wings or body to scrape off pollen as its tongue laps up the nectar. Many butterfly flowers also provide wide petals on which the butterfly perches while it drinks. Others in this category produce thick clusters of small flowers, like butterfly bush. The butterfly stands on one flower while lapping nectar from another, and gets pollen on its feet.

Investigations of the color vision and color preferences of butterflies carried out in the 1920s revealed preferences for blue, purple, and yellow. These experiments were performed in Europe with common sulfur butterflies. More recent research using the long-lived tropical *Heliconius* butterflies shows their strong preference for yellow. Since *Heliconius* butterflies feed on yellow pollen as well as transparent nectar, the bias toward yellow may be a useful innate response to their food. Both investigators reported that their butterflies showed almost no interest in green, brown, or white, and indeed these colors are almost unknown among butterfly-pollinated flowers.

A recent field study focussed on a small flowering tree named *Caesalpinia pulcherrima*, native to Mexico but planted in gardens throughout the New World tropics. It was once suggested that this tree is adapted partly for bird and partly for butterfly pollination. Investigation revealed, however, that the flowers of this tree open only at times of peak butterfly activity; in addition, their nectar spurs are so narrow that they can hardly accommodate a bird's beak, though they admit a butterfly's tongue easily. The tree is specifically adapted to butterfly pollinators. Its pollen grains, stuck together by thin gluey threads, cling to the wings of butterflies and are carried from tree to tree.

The *Heliconius* butterflies, with their unusual pollen-eating behavior, have been intensively studied. Their pollen-gathering, it turns out, is a skill that improves with practice; older individuals are better at it than young ones. Analysis of the pollen reveals that it contributes essential amino acids to their diet. The nutrient-rich diet of *Heliconius* butterflies may help to explain their long survival in the adult phase of life—two to three years, in contrast to the few months or weeks typical of most other species of butterflies.

A controversial recent study asserts that, in contrast to the foregoing statements, some butterflies are not benefactors but exploiters of flowering plants. The subject of this study, the European wood white butterfly, is a frequent visitor to wild violets and peas. Field captures, and assessment of pollen loads on the wings, showed that the butterflies carried only tiny amounts of pollen. The quantities transported were far too small to effect successful cross-pollination. Yet these flowers do appear to be butterfly-adapted, with their sweet scent, blue and yellow colors, daytime opening, and nectar spurs. The wood white butterfly may have evolved unusual skill at extracting nectar without carrying a payload of pollen. It does not follow that all butterflies lead the same exploitative existence.

Abroad while we sleep, moths are the logical recruits for partnership with night-flowering plants. Pale, sweet-scented flowers such as night-flowering cereus and evening primrose glimmer through the darkness to attract their visits. There are exceptions, of course: daytime flowers such as lilies and cleome may be visited by the hummingbird hawkmoth, which hangs on buzzing wings in the hot sunshine, licking nectar out of the flowers with its three-inch-long tongue. But the majority of moth-pollinated flowers bloom in evening or at night. Such flowers usually offer their nectar in long thin spurs suited to the long tongues of moths. Unlike butterfly-adapted flowers, they seldom offer a petal to land on. The moth must scrabble to hold onto the flower, or else hover in front of it on rapidly beating wings.

The frequency with which white or pale star-shaped flowers are found in moth-pollinated plants hints that even night-fliers like moths may use visual signals from their flowers. A dramatic tropical orchid *Angraecum sesquipedalis* combines a white star-shaped flower, a ten-inch-long nectar spur, and a delicious fragrance to attract a moth with the record tongue length of ten inches when unrolled! The pair provide a superb example of a co-evolved pollination partnership.

An equally tight relationship exists between the desert yucca plant and its moth pollinator, but here the partnership has overtones of infanticide. Yucca plants are pollinated by a small female moth that lays her eggs, a few at a time, in the depths of the flowers. When these eggs hatch, the tiny caterpillars start to devour the nearby flower ovules which should become the plant's offspring. If this infanticide were complete, the yucca would long since have become extinct. But the flowers produce enough ovules to feed their hungry guests and still have some left over. Careful calculation has shown that each caterpillar needs to consume about twenty ovules to reach maturity. On the average, six caterpillars live in each flower. Thus the caterpillar food quota is one hundred and twenty ovules per flower. The yucca flower actually produces, on the average, about two hundred ovules. The yucca plant can thus afford to sacrifice some of its young to serve as food for its pollinator's young and still have enough left to perpetuate itself.

Studies on the foraging behavior of the hawkmoths active by day suggest that they are as sensitive as honeybees to visual guidemarks on flowers. In an ingenious experiment they were offered cut-paper "flowers" placed behind a sheet of glass. Some of the paper flowers were streaked with guidemarks pointing toward their centers, and others had randomly oriented marks. The moths were allowed to feed briefly on a sugar-and-water mixture, and then fly toward the glass. As their long tongues unrolled and lapped against the glass they left sticky little dabs of sugar solution on it. When sprayed with red lead and heated, the nearly invisible sugary tongue-strikes showed up in color. A significant majority surrounded just those flowers centers indicated by orderly guidemarks.

Since these hawkmoths forage by day it is not too surprising that they should use their eyes. But the assumption has long been made that night-flying moths depend almost exclusively on fragrance to locate their flowers. Research with night-flying hawkmoths failed to support this assumption. When fragrant flowers were enclosed in glass tubes open at the top and were offered to night fliers in dim light, the moths flew right at the flowers themselves. They banged against the glass and failed to move upward to exploit the opening at the top from which the alluring scent diffused. In another experiment, moths that had been trained to feed from scentless artificial flowers were then offered real, sweet-scented, nectar-providing tobacco flowers as well. The moths stayed faithful to their scentless artificial food source for three nights before they decided to pay a visit to the flowering tobacco, even though this is one of their favorite plants in the wild.

Perhaps the studies on smell have failed to discover what the fragrances so common in moth-pollinated plants can do to manipulate these creatures. It seems improbable that so many of these plants would produce their heavy sweet smells for no one's benefit. And some practical evidence supports that assumption. Human moth collectors who want to attract their prey to a given spot smear the tree-trunks there with a sweet, sticky, smelly mixture of molasses, crushed bananas, and stale beer. The gummy mess usually succeeds in trapping huge numbers of moths in a few hours, and the smell is the only long-distance clue it emits. At least one investigator has been able to train moths to feed at specific flowers that differed only in odors. The whole story on fragrance is evidently not in yet as far as moths are concerned.

Pollen transfer by ants seemed like an absurd idea until recently. With their slick body surfaces ants appear to be very poorly adapted to carrying pollen grains. They frequently groom themselves, too, which would clean off any pollen that did manage to cling at first. On top of that, they are flightless, except for a very brief period in spring when winged males and females fly to mate. Aside from that nuptial flight, most ants are crawlers that stick close to home, so they would be less likely than any of the winged insects to cause cross-pollination.

In the past, when spotted on flowers, ants were labeled nectar thieves busy stealing the sweet sap meant for legitimate pollinators. As one authority on floral biology remarked, "If ever there was a scoundrel in the pollination drama, that role has been assigned to ants." But hot, dry environments poor in other pollinators strongly favor ants as pollination partners. Ants, after all, are widely distributed, highly social, and very numerous. These are characteristics that, when winged rivals are absent, help to balance out their bad points. In addition, their craving for sweets means that they will be likely to visit any nectar-secreting plant.

Floral features that have been considered peculiarly suitable to ant pollination include small overall size, with the flowers located close to the ground so that they are more accessible to ants; the production of small quantities of pollen, which will not trigger the grooming behavior; flowers that open only a few on each plant at a time, so that the ant forager must visit several separate plants and promote cross-pollination; and the tendency to grow in dense clusters, shortening the ants' commuting time. The plants for which ant pollination has, so far, been documented—knotweed in western Oregon, a little plant named *Orthocarpus* in the Pacific Northwest, and rupture-wort in England—are all small plants that fit these guidelines.

A commercially valuable crop widely grown in Africa, the Bambara groundnut, is another plant that seems to fit the ant-pollinated profile. When plots of this plant were experimentally caged to exclude ants, pod production was significantly lowered. A relative of our familiar peanut, the Bambara groundnut needs to sink its pods underground as the peanut does for best seed production. In the ant-deprived plots, many of the pods remained on the surface of the ground. The plants themselves were less vigorous and wilted earlier. Ant pollination may be critical for good seed-set in the Bambara groundnut, and its similarity to the peanut raises the question of what part ants play in the life of this even more important plant.

If a plant is normally pollinated by other insects but is also visited by robber ants, it frequently turns out to have devices to keep ants away from the floral nectar. The catchfly, a little wildflower found in Europe and North America, accomplishes this with bands of sticky secretion around its stem. Ants climb the stem to the flower and get trapped by the botanical tanglefoot, but winged (and presumably more legitimate) pollinators fly above it. Another device, seen in teasel flowers, acts like a medieval castle's moat: water collects in grooves at the base of each flower to repel or drown unwary ants.

A more controversial structure that may also be involved in keeping ants away from flowers is the extrafloral nectar gland. Discussed previously as a means of attracting an ant bodyguard, extrafloral nectaries may serve as well to bribe ants into staying on stems and leaves rather than robbing flowers. The plants that seem most likely to be bribing ant visitors in this way are the ones that begin production of extrafloral nectar at the same time as they open their flowers. Such precise timing suggests a method of ant control rather than, as B. J. D. Meeuse put it, a "regular salary" to ant bodyguards.

One group of plants that does coordinate extrafloral nectar secretion in this way with floral opening is the Asiatic species of *Impatiens.* Grown in the United States under a variety of names—patience plant, Patient Lucy, flowering balsam, New Guinea Impatiens—these plants all secret droplets of sugary nectar along the stems when they flower. Their extrafloral nectar is so rich that small children who have a Patient Lucy plant nearby have been known to lick the stems. Of more importance to the plant, the nectar glands are placed in such a way that ants climbing up the stems cannot avoid them. Stopping to drink from them usually keeps ants away from the flowers.

Usually, but not always. There is a peculiar twist to the *Impatiens* story. A few observers have noticed that ants can occasionally be found right in the flowers eagerly lapping floral nectar. It looks as if these plants have just attracted ants' attention by their extrafloral nectaries. The flowers are large, bright, and have long nectar spurs; they are not adapted for ant-pollination. In fact, it seems likely, considering the flowers' lack of fragrance and their habit of hanging on flexible flower stalks like fuchsia, that the Asian *Impatiens* are bird-pollinated like fuchsia. If this is so, why are ants encouraged to crawl up their stems and occasionally allowed into their flowers?

Most nectar-feeding birds—perhaps all—need to eat insects to supply protein in their diet. Could the Asiatic *Impatiens* be using their nectar flow, floral and extrafloral, to attract ants as extra tidbits for bird pollinators? The answer to this must await further field and laboratory investigation.

CHAPTER 7

How Doth the Little Busy Bee: Honeybees, Bumblebees, and Wasps as Pollinators

How doth the little busy bee
Improve each shining hour
And gather honey all the day
From every opening flower!
Isaac Watts, *Divine Songs*

BEES HAVE EARNED SUCH ACCLAIM AS POLLINATORS par excellence that "the birds and the bees" has become synonymous with sex and fertilization in general. (This seems to imply that a flock of birds and a hive of bees are necessary for our own reproduction, but most people eventually get this cleared up.) Bees are abundant and assiduous pollinators in the tropics, where many plants rely on them exclusively. Even in the temperate zones, they are the major pollinators of such common plants as goldenrod, aster, skunk-cabbage, and many others. And the dependence is mutual. Bees have evolved a nearly complete reliance on flowers for their food. With the exception of a recently documented case of carrion-eating behavior among South American bees, the diet of bees includes no meat at all, even in the larval stage of their lives. They depend on nectar for fluid and sugars, and on pollen for protein, fats, and vitamins. Their structures, their senses, and their behavior patterns have coevolved with floral foods.

Bees are well-designed by natural selection for the transport of pollen. They possess numerous branched or plumose hairs scattered over their bodies. As a bee walks over a flower, pollen grains cling to the hairs on the underside of its body, later to be brushed off onto other flowers. (Bumblebees, primarily dwellers in cold climates, have a very dense coat of hairs that helps to insulate them, while honeybees and other bees of tropical origin have thinner coats.)

An anatomical adaptation to pollen-feeding that is at its most advanced among honeybees is the presence of hollows on the hind legs, lined and edged with bristles. These hollows, known as pollen baskets or corbiculae, can be packed full with masses of pollen grains. Honeybees sometimes load their pollen baskets so full that they seem to have yellow pebbles stuck onto their hind legs.

A third structural adaptation to feeding on flowers is the honey stomach. This area of the gut, close behind the mouth, can expand to hold large quantities of nectar. It is this structure that allows bees to "gather honey all the day/From every opening flower." A bee can store the nectar gathered from several flowers here before returning to the hive to share the riches.

Eyesight, and the sense of smell, have evolved in bees in directions that reflect their flower-feeding behavior. Most research on the senses of bees has been done with the honeybee, as its economic value to us is so great. Most striking is the discovery made in the 1930s that honeybees prefer to visit objects with irregular, jagged outlines. When an array of cut-paper flowers, some with jagged silhouettes and others with uniform rounded or square edges, was offered to honeybees, they visited the irregularly shaped cutouts significantly more often. These shapes are reflected in nature by the jagged or lobed edges of flowers that bees love to visit.

The honeybees in this experiment preferred to alight on the outer edges of the cut-paper shapes. This suggested a conflict; the pollen and nectar of flowers are centrally located. Through their attraction to edges rather than centers, the bees would be likely to miss the mark. The problem is overcome by visual guidemarks that atract bees toward the centers of flowers. As in moth flowers, the marks may be dark streaks or spots, or occasionally streaks of white on a deeply colored flower.

As mentioned in Chapter 6, some flowers possess guidemarks invisible to our eyes. The grand old man of honeybee research, Karl von Frisch, published a number of photographs of common flowers in *The Dance Language and Orientation of Bees* that illustrate strikingly the difference between what bees see and what we see. These flowers were photographed using filters that exclude various wavelengths of light. When ultraviolet light was excluded, many of the flowers revealed streaks, spots, or target areas of ultraviolet reflective tissue marking the central area. Von Frisch had demonstrated earlier that bees do see ultraviolet as a distinct color. As if in invisible ink, the flower is drawing a roadmap to its treasure that only the chosen searchers can read. A common example is cinquefoil or Saint-John's-wort: it appears uniformly yellow to our eyes, but to bees it appears quite different, with dark streaks on the insides of the petals pointing directly to the center of the flower.

Some flowers are strongly ultraviolet reflective all over. This "bee-violet color" is attractive in itself to bees, and can account for a flower's popularity with them, as is the case in Oriental poppies. Other popular colors are blue, yellow, and pink.

The sense of smell as well as sight lures bees to their flowers. Von Frisch began his research on honeybees' color preferences by offering them tiny glass dishes of sugar water that rested on paper placemats of different colors. But he found that very few bees visited his restaurant. He then flavored the drinks with perfumes. The fragrances attractive to bees are varied, but tend to fall into the categories of spicy, flowery, or fruity—sweet smells that are pleasant to humans as well as to bees. Von Frisch succeeded in training bees to dine at dishes scented with any one of twenty-eight different floral odors. Putrid, pungent, or camphorlike smells failed to attract them. One such compound—skatole—can even repel bees when added to a pleasant fragrance.

Nearly all bee-dependent flowers are fragrant. In fact, Von Frisch found only three odorless species among all the many bee-pollinated plants he studied. They were the wild grape, red currant, and huckleberry. These plants may survive and set seed because they typically grow in dense stands. Accidental discovery by a bee of one flower in a large clump could lead, by investigation of the nearest neighbors, to pollination of the whole mass.

Bees also secrete scents that help in pollination. Many

species mark a rich food source with fragrances that target it for repeated visits by themselves and their hive-mates. And like Hansel and Gretel marking the path from their home with white pebbles, some bees are even able to blaze a trail back from a rich food source with tiny droplets of odorous secretions. Their hive-mates use these droplets to sniff their way back to the food. Both types of secretion qualify as pheromones, chemical secretions that serve as signals in animal communication.

Leaf-cutter bees, a wild species native to the tropics, use trail-marking pheromones extensively. Field observation shows that a scout bee from a leaf-cutter hive, once she discovers a rich food source, spits onto it a secretion from her mandibular glands. As she heads back to her hive, she pauses along the way to spit tiny quantities of the same secretion onto leaves and twigs. When she reenters her hive she runs about excitedly, knocking into other workers, and apparently thereby signals "There's food out there!" Workers recruited by her excitement dance leave the hive, and follow the scented trail she laid back to the food source.

Martin Lindauer, one of the generation of behavioral biologists inspired by von Frisch's work, describes a mean but ingenious trick that he played on leaf-cutter bees. He set their hive out directly across a small pond from a rich food source. A few scouts eventually found the food. They then made a beeline straight across the pond for their hive, trying fruitlessly to lay a trail of odor marks on the water surface as they flew. Their excitement dances recruited many other bees to exit from the hive, but without an odor trail almost none of them found the food. Their frustration was relieved when Lindauer strung a clothesline across the pond, with leafy twigs hung from it. The scouts now scented the rope bridge, and their recruits finally found the food!

An interesting aspect of odor trail-marking is that the trail fades slowly with the passage of time, as does the availability of floral food sources. A flower open and rich in nectar at 8:00 A.M. is not likely to be as rewarding to a bee at 4:00 P.M. So the odor trails must be up-to-date to be useful—and only if they are up-to-date are they detectable.

Not only leaf-cutter bees but even honeybees use fragrance as an aid to finding their favored plants. When a successful scout returns to a honeybee hive, she too performs a recruitment dance, which attracts close attention from her hive-mates. Essential oils from the flowers she visited cling to her body surface, and scent the nectar she carries in her honey crop. The watching bees sniff at her, and if hungry, solicit nectar from her, so that smell and taste both tell them the kind of flower she has been visiting. (If all the workers are temporarily full, they refuse to solicit nectar from the scout. She then stops dancing. The lack of interest tells her not to waste energy on signaling an unnecessary food source.)

Honeybee scouts mark rich flowers after feeding from them, using a pheromone from a gland on the abdomen known as Nasanov's gland. Occasionally, scouts are seen laying traces of this secretion on plants as they head back to the hive, although a formal odor trail like that of leaf-cutter bees is not common. When their Nasanov's glands are sealed shut with shellac, honeybees still forage on flowers. But upon return to their hives the sealed scouts are able to recruit only one-tenth the normal number of their hive-mates to visit the flowers.

The excitement dances of honeybees are considerably more sophisticated than those of other species of bees. When a scout honeybee has marked a rich food source with her Nasanov's gland, returned to her hive, and found by the readiness of hive-mates to accept her nectar that they are eager to share in the food source, she may perform either of two dances. The round dance, a simple circling, tells the other bees that the food lies close to their hive. Depending on the variety of honeybee, this dance says "Food within eighty meters" or "Food within one-hundred-twenty meters." The scout bee's floral fragrance tells the recruits what plant to settle on, when they get close to it.

If the returned scout performs a more complex figure-eight dance pattern known as the tail-waggle dance, she is signaling that food is at a greater distance. The direction in which it lies is indicated by the direction in which the "'waggle" part of this dance is done. The scout's speed in dancing, and the length of time she continues to dance, give her watchers an idea of how far off this food is and how rich a source it is.

Honeybees from one hive normally all feed on the same kind of flower for hours, even when other plants are also in bloom nearby. Bee specialists call this persistence floral constancy. It was known in the nineteenth century that a honeybee worker normally forages on only one sort of flowering plant on a trip; this helps to simplify the messages her recruits get. Floral constancy is an allegiance that is only sworn for a few hours, however, and only flowers that have very rich and abundant nectar usually win even this limited loyalty.

Floral constancy is clearly beneficial to the plant involved. Its pollen will be transferred to members of its own species. But at first glance it seems less clear how floral constancy benefits the bees. Increased efficiency seems to be the answer; by continuing to work on the same sort of flowers, the bees stay with a known shape. They can manipulate these flowers more easily as practice makes perfect.

For honeybees, floral constancy is a hive-wide phenomenon: in a given hour, all the workers from one hive will probably visit the same species of flower. Their choice is dictated largely by the enthusiasm with which the first few scouts of the day dance out their message about their finds.

A nectar-providing flower is "danced for" if its nectar combines high sugar concentration with a high rate of flow: quality plus quantity. If several returned scouts try to drum up business for several different kinds of flowers at once, the greatest number of recruits will be won by the most vibrant dancer. And she will be promoting a close, rich, and abundant nectar source. Thus the foraging activity of the hive as a whole is directed toward high-profit flowers.

Bumblebees, too, show floral constancy, but in an individualistic way. A successful bumblebee scout returns to her nest and stores her nectar and pollen there, to be eaten by her queen mother and her sisters. In that sense she is truly social, devoting her energies to the care of others. But she does not convey any information about her discovery to her hive-mates, even though she is storing her harvest beside theirs. She continues to visit the same flowers as long as they are rich in food, but she tells no one else about them. Bumblebee society has been compared to the old-fashioned cottage industry period in Western history, in which each artisan found his or her own raw materials, worked them, and marketed the product individually. Honeybee society has been compared to modern industrialism, in which certain workers gather raw material and others process it for low pay, with the profit almost all retained by the factory owner—the queen, in honeybee society.

The analogy is weakened, however, when we recognize that both bumblebees and honeybees practice communal twenty-four-hour child care. This is an amenity not widely available even in modern factories and not considered necessary in the day of cottage industries, when each family cared for its own. The comparison breaks down even further when we consider that reproduction in both bumblebees and honeybees is reserved for the ruling class. Only queens lay fertile eggs, as a rule. Their young are fed, cleaned, and fiercely protected by the older offspring, sterile older sisters who will die worn-out from their hard work long before the queen.

Bumblebees and honeybees differ in the degree of social enslavement to the queen mother. Bumblebee queens assert their privileged status by physically pushing the younger females around. Early experience of defeat in jaw-to-jaw wrestling convinces the younger bumblebees to knuckle under permanently to their mother and care for her offspring. But they are sterile only while she stays with them; her removal allows them to return to fertility. In addition, she forages for nectar and pollen, especially when her little colony has just begun. She is only a trifle heavier and longer than the worker bumblees.

In honeybees, the physical contrast between queen and sterile worker female is much more striking. The queen is a heavy, opulent figure, while her worker daughters are small, trim, and permanently sterile. The honeybee queen achieves control over her workers in a more subtle chemical manner than with bumblebees: a secretion from her mandibular glands, eagerly lapped up by worker bees, keeps them sterile and submissive. In this case, even if the queen mother is removed, the honeybees remain sterile. Without their treasured dictatress, the workers act dismayed; they soon rear a new queen to replace her. Their only act of violence against their tyrannical mother comes when her rate of egg-laying slows down in old age. When her fertility disappears, they lick her to death.

Honeybee colonies, because of their large size, are able to store enough food to last through the winter, and by clustering together during the cold they share enough body heat for most of the workers to survive along with their queen. These facts, combined with their preference for pollinating many economically valuable crops, have made honeybees the most popular subject of research on bee senses and foraging behavior. But recent attention to bumblebees has shown that they, too, have evolved their own set of fascinating responses to their world.

Probably the first association anyone has with bumblebees is the thought, "They're the ones that are really too heavy to fly, aren't they?" There is a kernel of truth in this old chestnut. If the wing muscles of bumblebees become colder than 30°C (86°F), they cannot beat fast enough to support the bee in the air. Yet field research on bumblebees in a bog in Maine found some bees in flight at air temperatures of 0°C (32°F). The paradox was resolved after tiny heat-sensitive recording devices or thermistors were implanted in bumblebee muscles. The thermistors revealed that on cold days the bees shivered hard enough to heat up their muscles to the necessary takeoff temperature. The colder the day, the longer the warm-up took. Any human runner preparing for a road race on a cold day has experienced the same thing, despite our supposedly warm blood. The American zoologist Bernd Heinrich, who carried out the study on bumblebee warm-up, is himself a talented long-distance runner; he won the over-forty division of the 1980 Boston Marathon. His avocation may have helped prime his professional interest in bumblebee energetics.

A warm-up through shivering expends a lot of energy. Thus it is not surprising that bumblebees visit low-energy flowers which provide dilute or scanty nectar only when the day is warm. At chilly temperatures, Heinrich found, they restrict their foraging to high-energy flowers whose rich nectar will give them enough energy to stay warm easily.

Bumblebees, like honeybees, gather pollen as well as nectar in order to feed their larvae. In gathering pollen from goldenrod—which flowers at a time of year when chilly mornings are common—they usually conserve energy by walking from one floret to another within a flower head or panicle. They fly from flower to flower, hovering on the wing in between, only when the nectar reward is high.

On most of the plants they visit, bumblebees appear to probe flowers at random, frequently returning to flowers they have just emptied. But this "random search" is not always the case, as evidenced by the foraging technique of bumblebees on three species of tall plants with narrow flower spikes. Bumblebees feeding at foxglove, delphinium, and monkshood, which all have tall, narrow spikes of flowers, seem to follow the rule of starting to forage at the base of the flower spike and moving upward. This pattern minimizes fruitless second visits to empty flowers.

In the case of foxglove, eighty percent of the observed bumblebee visits to flower spikes began at the bottom of the flower spike and went up. The foraging technique in this case benefits the plant as well as the pollinator. Sexual structures in foxglove flowers develop so that the oldest flowers, at the base of the spike, are female, while the newest ones, at the top, are male. When a bee alights at the base of a foxglove spike, thanks to floral constancy she carries in pollen from a neighboring foxglove. She leaves the pollen on the lower flowers, and acquires more higher up the spike, before flying off. Thus the foxglove's cross-pollination is facilitated.

Many native North American plants, such as goldenrod and aster, depend heavily on bumblebees for pollination, and compete with one another for their visits. The native flowering plants tend to bloom either early in the summer, such as blueberry bushes, or late, such as goldenrod. Measurements of nectar flow reveal that the early-season bloomers are rich nectar sources, while the late-season bloomers produce very scanty amounts.

This pattern of nectar flow, Heinrich suggests, is a reflection of the bumblebee life cycle. Only a few bumblebees survive the winter each year. These are generally fertile young queens who emerge from below-ground in spring to establish new colonies. But in spring and early summer, their numbers are low. The early-blooming native wildflowers must compete intensely for the pollination services of the few bumblebees, and do so by offering a rich reward of nectar. The sugary fluid gives the bees strength for the more arduous tasks of building nests and gathering pollen to feed larvae. Thus their numbers multiply. By late summer, there are lots of bumblebees. The late-blooming flowers need to offer very little in the line of nectar reward. The bees are now competing intensely with each other for scanty floral nectar on each plant. Co-evolution allows the flowering plants of late summer to conserve their energy by cutting the pollinator payroll, now that the job market is flooded with eager workers.

Even within one part of a growing season, accommodation to pollinators is visible among native North American wildflowers. Field observations have suggested that they take turns at blooming, and by this staggering of flowering peaks reduce their rivalry for the pollinators. In related plants that have such staggered, or offset, blooming periods fairly close together, the earlier-blooming plant provides more nectar than the latecomer. In this relationship, the generous early bloomer teaches the bees to expect a continued rich reward so that they will continue to visit even the miserly late bloomer.

A native bog plant that achieves maximum services for its flowers, with no payoff at all to the bumblebee, is an orchid known as the grass pink. Why should the bees visit its unrewarding blossoms? Three factors combine to make this exploitation possible. The flowers are unscented and therefore are hard for bees to remember as a poor employer. The flowers are variable in color, ranging from white to deep pink; this makes them hard to learn visually as well. And the flowers of grass pink strongly resemble the flowers of rose pogonia, a bog orchid that does release nectar. The grass pink can fool the bees after the rose pogonia has fed them.

Monkshood is a Western flowering plant that, like New England bog plants, has a strong dependence on bumblebees. Its flowers, shaped like deep pointed hoods, grow on tall spikes similar to foxglove or dephinium. Some populations of monkshood have nectar glands at the front of each flower while others have them deep in the throat of the flower. The monkshood plants with easily accessible nectar attract both long- and short-tongued bumblebees. The nectar output of these flowers is significantly greater than in the shallower flowers, so that the bumblebees are better paid for their visits to these than to the more easily worked flowers—and so they keep coming.

Milkweed, distributed across North America, has a peculiar pollination relationship with bumblebees. Milkweed flowers produce pollen in two bundles or pollinia connected by a little bridge, the corpusculum. When a bee steps on the flower, the pollinia cling to its feet. Field research shows that in many cases milkweed pollinia are so bulky that they make the bees stumble over their heavy feet and fall off the flowers! Most of the bees then fly some distance away before they settle on another milkweed. This promotes outbreeding, because milkweed plants send out underground stems that produce nearby clones of the central plant. Near milkweed neighbors are likely to be genetically identical, while those at a distance are likely to be dissimilar.

The pollinia sometimes get the bees tangled up in the milkweed flowers, and in disentangling themselves they may tear off claws, feet, or even their lower legs. The heavy pollinia also demand that the bees spend more energy when they carry them just to stay aflight. It seems, in fact, as if bumblebees are in for nothing but trouble if they visit milkweed flowers. Why do they keep doing it? The secret lies in the high pay scale for this hazardous job. Milkweed flowers produce copious amounts of rich nectar. Although the risks of working a milkweed flower are high, the rewards compensate for it, like the situation with human workers who help build skyscrapers.

Orchids, like milkweeds, offer their pollen in pellets or pollinia, and like milkweeds they are mainly dependent on bees for pollination. But the orchid family includes flowers that offer even stranger incentives to their bee pollinators. One tropical species provides neither nectar nor pollen but an abundant fragrant oil in the surface cells on its petals. Male wild bees visit these flowers to scrape at the petals. They collect and smear the fragrant oil on themselves and, in the process, pick up the orchid pollinia. They visit several such orchids, thus effecting cross-pollination, and then perch on twigs and wait for their acquired fragrance to attract females. The scented oil works so well as a come-on for these males that it reminds us of the old TV ad "Brylcream, a little dab'll do ya/Use more—only if you dare/Brylcream, the girls'll all pursue ya/They love to run their fingers through your hair."

In a little tropical orchid, *Oncidium planilabre*, even this oily reward is not available, nor is nectar or edible pollen. Instead, the tiny yellow-and-brown flowers of this orchid, poised on thin wiry stalks, tremble in every breeze, suggesting a swarm of large bees. Male wild bees in the vicinity take them for rivals, and attack them as they would one another, ramming the flowers with their heads. Pollinia stick to their foreheads, and are transferred later to the next flower that seems to be looking cross-eyed at them. This orchid is said to be practicing pseudoaggression. Even though the flowers are no match for the bees in fight, there is no question that they are winning their private battle for survival.

A special case of orchid deception involves wasps. The little European and North African orchid *Ophrys* supplies no nectar, pollen, or fragrant oil, yet it is regularly visited by wasps. Its flowers are of a peculiar deep metallic blue and green, with an odd wasp-waisted shape. As long ago as 1916, the French naturalist A. Pouyanne theorized that the wasps were being deceived by the flower into thinking it was another wasp, and recent research has confirmed this idea. Male wasps approach and clasp *Ophrys* flowers, attempting to mate with them, so that these plants succeed in pollen transfer through pseudocopulation.

Other insects, too, believe that *Ophrys* flowers are wasps. When several of these orchids are cut and pinned onto flowers usually visited by bees, the number of bee visits to such decorated spikes falls drastically. If we recall that wasps are predators that are not above attacking a bee to make a meal, this avoidance is understandable.

Exploitation of bees by man has a history as old as the Pyramids. The honeybee-keepers of ancient Egypt floated hives on reed boats along the Nile for the successive pollination of crops on the banks. A major commercial use of honeybees as crop pollinators now is in apple orchards, where hives are usually placed at least every quarter-mile throughout the trees to ensure good fruit-set. But while honeybees work well for many crops, there are certain plants on which they work poorly. They are small insects, so they lack the weight to force open heavy, stiff flowers. Their tongues are short, so they cannot lick the nectar from deep flowers, and as a result refuse to visit them often. Bumblebees, in contrast, are large, heavy-bodied insects with longer tongues. They succeed far better than honeybees as pollinators of red clover, an important cattle-fodder plant.

In the case of alfalfa, another important fodder plant for cattle and horses alike, honeybees refuse to visit the stiff, spring-trap-like flowers more than once. Several species of wild bees are therefore crucial to alfalfa's success. The best-known of these is the northern leaf-cutter bee. This strong little bee, an immigrant to the United States, gets its common name from its habit of chewing neat disks out of the edges of leaves. It uses the disks to line small holes in tree-trunks, or insect tunnels in the ground, for its nest. Human farmers can attract leaf-cutter bees to alfalfa fields by providing artificial domiciles or bee-houses; some have been made of stacked soda straws, others of layers of wooden planks grooved to form tubular dwellings between the layers. These are usually well-acccptcd as housing. The bee apartments have one drawback if they are all the same color. Individual leaf-cutter bees have a hard time locating their own nests at the end of the day, like tired human commuters returning to a tract housing development. To save the bees trouble, builders of these artificial domiciles now paint blocks of them different colors.

Another wild bee, the alkali bee, also pollinates alfalfa. It digs burrows for nests in bare soil. Alfalfa growers have dug up chunks of soil riddled with alkali bee tunnels and transported these, at considerable expense, to the edges of their fields. The resettled bees, it is hoped, get down to business and start pollination in a day or two. In a variation on this technique, building sites for alkali-bee housing have been made by digging out the soil at the edge of a field, and then backfilling with the alkaline earth they prefer.

Subsidized housing has also been provided for bumblebees in order to promote the pollination of red clover and other crops. Underground dugouts with wooden lids, or small wooden boxes lined with upholstery stuffing, have been used. These manmade dwellings meet with approval by their prospective tenants; in one study, nearly three-fourths of all the nest boxes were found to be occupied by breeding female bumblebees.

Despite their willingness to accept the favors we offer them, bees remain essentially independent of humanity. Even in the case of the supposedly domesticated honeybee, a swarm can and often does leave its manmade hive to seek food and shelter in the wild. Rather than being our slaves, they are temporary partners with us in the task of exploiting flowering plants for food. That very independence is part of their perennial fascination.

CHAPTER 8

Hummers, Creepers, and Fly-by-Nights: Warm-Blooded Pollinators

CERTAIN BIRDS AND MAMMALS HAVE FORMED POLLINA-tion relationships with flowering plants. In contrast with insects, these warm-blooded animals place great demands on the flowering plants that form partnerships with them. Their higher rate of metabolism—their warm-bloodedness—means that they must burn more food per minute than insects. As a result, the flowers they patronize must produce a richer or more abundant nectar. The greater size and strength of birds and mammals also means that they can easily inflict damage while they feed on a flower. Plants that entice birds and mammals must offer flowers that are larger, sturdier, or in other ways better guarded against injury than the insect-pollinated flowers. Beyond

these general guidelines, the good color vision and poor
sense of smell of birds impose a different set of require-
ments on their floral partners than do the poor color vision
and excellent sense of smell of most mammals.

Among the birds, hummingbirds, sunbirds, honey-
creepers, and flowerpeckers have evolved pollination syndromes. Of these, the best-known are the hummingbirds.
More than three hundred species of hummingbirds have
been recognized in the New World. Their closest living
relative, the swift, is a small insect-eating bird. Ancestors
of the hummingbirds were probably insect-eaters themselves, and modern hummingbirds still eat insects, but the
mainstay of their diet is nectar drunk while hovering on
the wing before a flower.

Plants that have coevolved with hummingbirds produce flowers that are odorless, bright-colored, and rich in
nectar. They lack guidemarks; the nectar is usually stored
in one or more deep spurs into which the bird's beak fits
like a key into a lock. Landing platform petals, too, are
absent; since hummingbirds feed on the wing in midair,
such petals are unnecessary equipment for the flowers. The
base of the flower may be thick and sturdy, or it may be
sheathed with a heavy protective calyx. These structures
protect the ovules of the flower against accidental damage
from the sharp beaks of their pollinators. Fuchsia, wild
columbine, and a few other bird-pollinated plants offer a
different protective feature: the flower hangs by a thin,
yielding stalk. When birds feed from these flowers, the
impact of their beaks makes the flower swing away and
thus escape damage.

Most hummingbird flowers found in North America are
reddish-orange or red. This has led many biologists to

assume that the birds have an innate preference for these colors. Yet many hummingbird plants in the Central and South America tropics have pastel-colored flowers. Why should hummingbirds accept these flowers in the tropical forest but opt for only red or orange flowers in the North? The American biologists Karen and Verne Grant, who conducted an extensive study of hummingbirds and their flowers, believe that the solution to this paradox lies in the birds' migratory behavior. Most North American hummingbirds spend winter in the tropics, returning yearly in the spring to their northern breeding sites. They cover great distances and encounter a great variety of flowering plants on the journey. If at some point in the past a genetic accident caused several North American plants suitable for hummingbird pollination to produce red flowers, they would all benefit from the birds' ability to associate the color with a food reward. Once the birds learned that this color was the signal for a rich nectar flow, they would be likely to visit unfamiliar flowers that gave the same signal. Red flower color would become an advertising logo. And in fact laboratory experiments with hummingbirds show that they can learn to associate almost any color—even ultraviolet—with a food reward.

Cardinal flower, a beautiful deep red, is one of the common North American plants dependent on hummingbirds. Field study has shown that although most cardinal flowers release abundant nectar, some populations release none at all. Yet even these unrewarding cardinal flowers receive regular hummingbird visits that pollinate their flowers. In a parallel to the grass pink/rose pogonia relationship for bumblebees, the nectar-free cardinal flowers are able to exploit their pollinators by close resemblance to flowering plants nearby that do produce nectar. The models for these stingy plants may be nectar-rich cardinal flowers, or even flowers of other species that like themselves are red and scentless.

This is essentially a case of Batesian mimicry, and as we saw in Chapter 4, such mimicry to be successful demands that the mimics be rare compared to the models. This is true of the nectar-free cardinal flowers. The numerical relationship of these to normal ones ensures that most of the time a bird visiting red tubular flowers will be rewarded. The positive reinforcement encourages visits to similar flowers. The few nectar-free flowers can then be mentally dismissed as bad luck. Furthermore, these deceptive cardinal flowers have evolved a peak of blooming that coincides with the northern migration of flocks of young, inexperienced hummingbirds. If plants could speak, these flowers would probably be heard murmuring P. T. Barnum's motto, "There's a sucker born every minute."

Hummingbird flowers, which open by day, generally close up at night, or if they stay open, reduce their nectar flow at night. But the basal metabolism of hummingbirds, measured by daytime, has led to predictions that a twelve-hour span of time without food would exhaust their energy reserves and leave them too weak to begin work in the morning. And yet they do appear, energetic as ever, each day. This discrepancy is based on the assumption that the high rate of hummingbird metabolism measured during the day continues during the night. In fact, hummingbird metabolism not only slows down during sleep, in many species it actually drops so low that the birds are as lethargic as a hibernating chipmunk. They can be taken for dead. In one species, the blue-throated hummingbird, the heartbeat rate drops from over 1,000 beats/minute during the day to about 30 beats/minute at night. This nocturnal torpor or lethargy allows the bird to conserve its energy resources so it can return to full vitality in the morning.

The deep, tubular form of most hummingbird flowers excludes most other pollinators. In many cases, indeed, the flower is so deep that only an unusually long-beaked hummingbird can reach into the nectar spur. Flowers of this sort are specialists, deeply dependent on hummingbirds, and they produce more nectar than a shallower, more generalized flower, does. But the relationship does not always work perfectly. A field study found that while the deeper flowers in a Central American community were mostly visited and pollinated by long-beaked hummingbirds, shorter-beaked species were not above stealing from them by boring holes through the base of the flower and drinking the nectar that oozed out! One small bird in the area— not even a hummingbird—led this life of a professional thief. It never pollinated any of the flowers, but drilled through their bases and drank the nectar from the wounds.

Such "floral larceny," to use the expression of another field investigator, is seen in other animals as well. One tropical, hibiscus-like shrub depends on the hermit hummingbird, a long-beaked species, for pollination. Short-beaked hummingbirds in the area occasionally deprive the legitimate pollinators of their reward by drilling the bases of the flowers, but the real culprits in this crime are our old friends the leaf-cutter bees. They eat holes through the flowers, drink their nectar without pollinating them, and—as if to add insult to injury—mark them with a pheromone that recruits more bees to the injured flowers. In addition they frequently form insolent groups that drive away the hermit hummingbirds from their flowers.

In a few cases, visits from bees may be helpful to so-called hummingbird flowers. Field studies conducted on hummingbirds at a research station in the Rocky Mountains of Colorado include one that examined pollination of a small wild columbine. This flower is bright red and yellow, odorless, and has deep nectar spurs. It fits the classical profile of a North American hummingbird flower. Yet three species of bumblebee also visit these columbines regularly. The nectar, in its deep spurs, isn't accessible to these bees, but the pollen is. The bumblebees gather the pollen, and while most of it is fed to their young, some of it is successfully transferred between columbines, enhancing the cross-fertilizing activity of the "legitimate" pollinators.

Another species of wild columbine in the same area has fragrant blue or white flowers; it is also visited by bumblebees. Cross-transfer of pollen between this species and the red-and-yellow species would be undesirable, since hybrids between species are often sterile. What prevents such accidental hybridizing, since both species are visited by bees? The ecological preferences of each columbine help to keep them distinct. The red-and-yellow species grows along shaded stream banks, while the blue or white fragrant columbine favors open, drier areas. They each occupy a different realm and are seldom visited by the same bees.

Another Rocky Mountain study examined the question of how two perennial species of flowers, both dependent primarily on hummingbirds, avoid competition and also possible hybridization. The plants turn out to use an elegantly simple method called sequential flowering: one species begins to bloom only after the other is nearly through with blooming. During the brief period of overlap, hummingbirds do visit both species; careful observation reveals that at this time each perennial shows a sharply reduced seed-set, the penalty for wrong-way pollen transfers.

A similar process of sequential flowering has been documented for hummingbird flowers in a Costa Rican forest. Here, ten of the species of flowering plants that coexist are dependent on one species of hummingbird for pollination. Here too the plants have staggered periods of bloom, so that at any one time only from one to at most three species are in bloom. The flowering plants here, as in the less complex Rocky Mountain community, have partitioned their pollinator's services, and thereby avoid competition and sterile hybridization.

Field study in a very different tropical habitat—not mature forest, but cleared land gradually returning to forest—reveals a similar pattern of bloom among the hummingbird flowers there. Such a habitat is termed successional, since year by year different kinds of plants will succeed one another until the mature forest is reestablished. It is truly a community in a state of flux. Yet even here the hummingbird flowers show sequential blooming seasons, again reducing competition and hybridization among the species. An ecologist is no more surprised that the flowering plants in a stable community stagger their blooming peaks than a sociologist is surprised that long-term neighbors in a human community watch each other's house on vacations. But a sociologist might well be taken aback to detect such neighborly cooperation among transient residents in a rooming-house, and ecologically the staggering of bloom peaks in a successional community represents just such a surprising act of cooperation.

In another tropical field study of hummingbird flowers, the investigator measured nectar flow from five of the most important species in one community. Within each species, nectar flow varied tremendously—some plants secreted great quantities, others very little or none. This pattern has been labeled one of "bonanzas and blanks." The feeding birds encounter many flowers with little or no nectar—the blanks—and only a few with abundant nectar—the bonanzas. Why don't they become discouraged and abandon the plants?

The biologist conducting this study noticed a similarity in this pattern of flowering to a technique used in operant conditioning by the psychologist B. F. Skinner. Skinner has trained pigeons to perform tasks reliably, without consistent reward, by means of what he calls a variable reinforcement schedule. The essence of this schedule is that a reward appears immediately after the task only *sometimes*, and unpredictably. Variable reinforcement, surprisingly enough, is even more powerful in teaching new behavior patterns than a consistent schedule of reward.

Plants that utilize the bonanza–blank pattern of individualized nectar flow reap several rewards. They save on energy, since the output of sugary nectar represents a loss of energy for the plants. Even more important for reproductive purposes, they promote hummingbird movements between flowers, as the birds search for bonanzas. Greater rates of movement promote higher pollen transfer from one plant to another—a higher rate of outbreeding.

Many flowering plants in the New World tropics have

coevolved with hummingbirds. What happens when the birds are no longer available as a partner in pollination? This separation can happen when a flowering plant goes island-hopping. Its seeds may travel, via man or fruit-eating bird, to an island remote from its original habitat, there to sprout and grow. But nothing dictates that the appropriate hummingbirds shall make the trip with it, and usually they do not.

Studies on such island-hopping in the West Indies have focussed on two flowering plants: *Justicia*, a moderate nectar producer with shallow flowers visited by short-beaked hummingbirds, and *Mandevilla*, a heavy nectar producer with deep flowers dependent on long-beaked hummingbirds. Both plants, with their respective pollinating hummingbirds, are found on Trinidad. Both plants are also found on Tobago, an island some distance away, but Tobago is home to far fewer hummingbirds than the bigger island. The long-beaked species on which *Mandevilla* is most dependent in Trinidad is totally absent from Tobago.

The investigators used an ingenious technique—mixing fluorescent dye powder with the pollen of both plants—to trace pollen movements in both locations. They found that in Trinidad both kinds of plant successfully dispersed their pollen. *Justicia* did well on Tobago, too, but the *Mandevilla* failed miserably. Its flowers received only sporadic visits from hummingbirds. Most of these were short-beaked visitors to many other species of plants. *Mandevilla* pollen dispersal and consequent seed-set were far lower on Tobago. The reduction in pollinators caused by island-hopping hit harder at the specialist *Mandevilla* than at the generalist *Justicia*. (Interestingly enough, though, nectar flow from *Justicia* is measurably higher in the Tobago population than in its Trinidad population. Apparently even generalists must increase their attractions to ensure pollination in a sparse island environment.)

Sunbirds, a family of small nectar-eating birds, pollinate bird-dependent flowers in Africa. Like hummingbirds, they visit bright, odorless flowers with rich nectar. Field study, and careful analysis of their feeding behavior, suggest that they follow several unwritten rules. The most obvious is the test-probe rule: after landing on a flowering plant, the sunbird probes one flower, and if it finds that the nectar content of this flower is low, it immediately flies to another plant. The rule here seems to be "Don't waste your time on clunkers."

Male sunbirds establish territories that they protect against other sunbirds. If successful in territorial defense, they achieve nearly exclusive feeding rights on a specific patch of flowering plants. In time, a territory holder learns the exact layout of the plants in his space, and within it he follows the rule of avoiding flowers he visited earlier that day. This allows him to save energy, since he will not try to feed from flowers he has already drained of their nectar.

It seems as if territorial defense might be bad for the plants in the defended area, since pollen dispersal is thereby restricted. Occasional interlopers get into every territory, though, and these may give just the needed amount of pollination across territorial boundary lines.

Among mammals, the most important pollinators are flower-feeding bats. These are in the minority among bats: the majority are meat, insect, or fruit eaters. But certain bat species in the East Indies and Australia are firmly committed to flower feeding, and a few species have been discovered in North America as well.

Flower-feeding bats, all fairly small, drink nectar and eat pollen. The plants they frequent show a number of adaptations to the furry pollinators. The flowers generally open after sundown since, as one investigator remarked, bats "work the night shift." The flowers produce abundant nectar to meet the high demands for energy of a warm-blooded flying animal. Last but not least, the flowers often dangle on very long stalks from the main mass of the plant, especially if it is a tree or vine. This peculiar arrangement, dubbed flagelliflory by pollination ecologists, allows the bats to feed from the flowers without entangling their delicate wings in twigs that could tear them. Other plants attract their bat pollinators by blooming before any leaves are present, or by elevating their blooms above the foliage mass on thick stalks.

Bat-dependent flowers give off a strange odor best described as "batty"—that is, they smell like a place that harbors a flock of bats. This odor exploits the sociability of flower-feeding bats. They are highly gregarious, and the smell suggests that other bats are in the neighborhood, which encourages them to alight on the plant. It is an olfactory parallel to the visual decoy flies offered by Venus's slipper orchids to the flies that pollinate them.

Bats have poor vision, and also appear to be unable to discriminate colors. Many bat-dependent flowers, in response, have coevolved to be white, pale, or greenish-drab in color. The pale flowers alert the bats' visual sense better than any bright color could at night. The greenish-drab flowers seem to have thrown in the sponge on visual attraction. They save energy by not creating any colors, and rely on odor as a bat-attractant.

In one white-flowering African tree that is dependent on bats for pollination, field study showed that bats spend only a brief time on each flower. Since the dark-furred bats are very conspicuous against the white flowers, and since nocturnal predators in the area—owls and genet cats—hunt by sight, coevolution may well dictate that bats feeding on these flowers survive by using a hit-and-run feeding pattern.

Long thought to be restricted to the Old World, bat pollination has recently been documented in the American Southwest. Nearly thirty years ago one pollination ecologist speculated that the saguaro cactus found in Mexico and Arizona might turn out to be bat-pollinated, since its flowers fit the bat-dependent syndrome. Photographs published in 1969 confirmed the speculation: they showed bats feeding on, and in, the large cactus flowers. These bats are small but strong fliers, capable of hovering on the wing while feeding—an important skill, since saguaro flowers offer no landing platform. Their little heads have long pointed muzzles, and their tongues are long and flexible. As in hummingbirds, long-tongued bumblebees, moths, and butterflies, the long tongue permits the bats to probe deep flowers for their nectar. The tongue in some cases is two-thirds the length of the bat's body. For a human parallel, imagine a six-foot-tall man with a tongue four feet long!

The tongue in flower-feeding bats is an erectile organ. When they are not feeding it is soft and small, and takes up very little room in the mouth. Many flower-feeding bats also have a fine brush of fleshy projections at the tip of their tongues, which help to lap up nectar. Flower-feeding bats, unlike most other bats, are nearly toothless; some are completely so. Their anatomy tightly binds their fate to the fate of the plants that feed them.

Field studies in the American Southwest have also shown that flower-feeding bats are the major pollinators for agave, the beautiful desert plants that may take fifty years to bloom. To trace the bats' movement at night one clever investigator tagged them with tiny glowing light bulbs that turned the bats into furry fireflies. She also extended her observations with ultrasound detectors placed on the agaves; these machines detected and recorded the high-pitched squeaks typical of bat communication.

She discovered that flower-feeding bats circle the agaves in flocks; one by one, each member of the flock takes a turn to swoop down, visits a cluster of flowers on the huge flower stalk, and then rejoins the group. After twenty minutes or so, the flock quits feeding and flies to a nearby resting site. By this time, their little bellies are bulging with nectar and their soft fur is coated with sticky yellow pollen. (It may have been one of these bats that a mammalogist in the Southwest once took for a species new to science—a bright yellow bat.) There seems to be no debate about the decision to stop feeding—one bat flies off and the rest follow immediately. Such group cohesiveness saves time that might be otherwise wasted. Efficiency is essential, since the bats work so hard for their food; hovering is one of the most exhausting forms of flight. But a good-sized agave will provide a rich enough payoff to satiate a flock of twenty or more bats.

What about other mammals? As long ago as the 1930s,

the European biologist Otto Porsch suggested that non-flying mammals might play an important part in the pollination of tropical trees and shrubs. His theory, largely ignored until recently, has now received support from field observations by several biologists. Species from at least three groups of nonflying mammals have been implicated as pollinators for certain flowers: lemurs in Madagascar, climbing marsupials in Australia, and mice in South Africa.

The plants that these mammals pollinate usually have large flowers that produce copious nectar. The petals are often white or pale yellow, and the blossoms frequently dangle from the long, tail-like flower stalks). The flowers dead-rat tree of Africa (its second name derives from the appearance of the fruits, approximately rat-sized, which dangle from the long, tail-like flower stalks). The flowers of many trees dependent on nonflying mammals, in fact, look as if they should be bat-dependent. But they grow in regions where flower-feeding bats are rare or unknown.

Some biologists believe that lemurs and marsupials pre-dated bats as pollinators, but were later out-competed by the more mobile bats. Under this theory, nonflying mammals should be important as pollinators in those regions of the world where flower-feeding bats failed to establish a foothold.

One such region is the island of Madagascar, off the east coast of Africa. Forests on the island are rich in flowering trees and in climbing lemurs, but harbor very few flower-feeding bats. The large lemurs active by day are destructive to flowers, but the small nocturnal lemurs—some barely bigger than squirrels—lick nectar from blossoms without damaging them, and carry pollen on their soft fur. The abundant nectar flow of their preferred flowers gives them a substantial amount of sugar and fluid.

In parts of Australia where flowering trees and shrubs abound but flower-feeding bats are rare, climbing marsupials have attracted attention as pollinators.

The best-studied climbing marsupial, the honey possum, shows several adaptations to a life of flower-feeding. Like the flower-feeding bats, this small mammal has a long pointed muzzle and a long, flexible tongue with tiny fleshy projections at the tip and along the upper surface. Nearly toothless, it depends heavily on nectar and pollen for food. Honey possums creep up onto flowering shrubs and trees, and when dissected their stomach contents usually consist only of nectar and pollen.

The fact that honey possums and other marsupial pollinators are common in Australia but nowhere else has generated some debate. A recent theory suggests that marsupials were once abundant throughout the world. Marsupials eject the embryos from the uterus very early and supply them with milk to grow on in an external pouch. During the era when flowering plants began their great success on land, small marsupials might have been their major pollinators. But as time passed, two rival groups arose to out-compete the pollinating marsupials. Social bees, armed with wings and stings, were bound to be faster than the climbing marsupials in floral visits. And placental mammals, including bats, also came into existence. These mammals retain their embryos in the uterus until they are far more advanced than the marsupials' embryos, and thus better able to survive after birth. The placentals soon outnumbered the marsupials by their sheer reproductive efficiency. Only in Australia, an island continent where social bees and most placental mammals never penetrated, could the marsupial pollinators continue to succeed.

A problem with this theory is that bats *did* reach Australia millions of years ago. Why, then, did the flower-feeding bats fail to oust the marsupial pollinators? The answer may lie in the bats' preference mentioned earlier for flowers that are either suspended well below the foliage or stuck up well above it. Many flowering trees in Australia —the banksias and bottlebrush trees, among others—have flowers right on the branches, among the twigs. These are almost inaccessible to bats, which are wary of tangling their wings, but are readily accessible to small climbing marsupials.

A separate question in regard to marsupial pollinators is whether any plants show special adaptations to these visitors, paralleling the adaptations some plants show to birds or bats. Some of the banksia trees, in the opinion of certain Australian biologists, do show just such coevolved adaptations. Their followers are dull in color, reflecting the marsupial lack of color vision, and are strongly scented with a heavy, animal-like odor that would attract the attention of marsupials with their excellent sense of smell. These flowers open at night, when the honey possums and other marsupials are active; and they grow low down on the plants, where the least climbing would be necessary. In addition the pistils, the projecting tips of the ovaries, are wiry and curved in these banksia flowers, so that they would press against a possum's furry head, coating it with pollen.

But other Australian biologists, skeptical of the idea, have argued that banksias are just as well adapted for bird as for marsupial pollination. Many flower-feeding Australian birds become active at early dawn; nocturnal nectar flow may just be "filling the feeders" for these birds. Some of the banksias with curved pistils are odorless, a characteristic associated with bird pollination, while some that have the heavy animal-like odor have straight pistils. And at least one species of banksia that is frequently visited and pollinated by honey possums has bright scarlet, high-placed, odorless flowers with straight pistils! The controversy is still unresolved.

Most recently reported, and at first glance least likely of all, is the case of pollination by mice. The report comes from South Africa, where, as on Madagascar, flower-feeding bats are rare or absent. The plants involved, natives of South Africa, belong to a family known as the proteas. Most proteas are tall plants with beautiful artichoke-like flowers pollinated by beetles or birds. But a few produce large, heavy, yeasty-smelling flowers that hang close to the ground. These proteas are regularly visited by mice. The little rodents gnaw the fleshy reddish leaves at the bases of the flowers, and stick their pointed muzzles into the flowers to lap up nectar as well. Pollen clings to their furry foreheads and is transferred later to other protea flowers. A feature of these proteas that may well represent an adaptation to these unusual pollinators are extremely thick, short flower stalks strong enough to support a mouse on the flower.

It would be remiss to leave out the most recent addition to the long line of animal pollinators—human beings. Carved stone reliefs from ancient Babylon show hand pollination of the date palm. Even today, from the hand pollination of *Vanilla* orchids to the mechanical rattling of greenhouse tomatoes in flower, human involvement in pollination is substantial. Most often it has become essential, as the changes of cultivation take a plant away from its natural pollinator. Human owners must carry out their end of the bargain, then, as surrogate partners.

Children, Go Where I Send Thee: Animal-Aided Plant Dispersal

Children, go where I send thee,
How shall I send thee?
I'm goin' to send thee one by one,
One for the little bitty baby
Born, born, born in Bethlehem.
AFRO-AMERICAN SPIRITUAL

SUCCESS IN POLLINATION LEADS TO THE SETTING OF seeds. The seeds are the plant's main link to the future; their formation is the reward for all the stratagems seen in pollination syndromes. Yet their formation creates a new problem for the plants: how can they establish a life for their offspring without depriving themselves? The soil in which a plant grows is its source of minerals and water, and those resources are limited. If offspring grow nearby, they will compete with the parent plant for these limited resources. It is biologically in the parent plant's best interests, then, to get its offspring to settle down somewhere else.

Moving away from the parent plant has certain advantages for the offspring, too. They thereby avoid competition with a mature and well-established rival. They also avoid the crowding that would make them conspicuous to their enemies. Separation makes it measurably less likely that any young plant will attract the attention of a hungry herbivore; in one study, elm seedlings beneath the parent tree were five hundred times more likely to suffer insect damage than those sprouting at a distance.

Certain plants, such as dandelions and elm and maple trees, cast their light seeds to drift to the four winds. Others, like jewelweed and violets, have evolved an explosive mechanism to catapult the seeds out of their pods. But many plants have become dependent on animal partners not only for pollination but also for the successful dispersal of their seeds.

As in pollination ecology, biologists once assumed that wind dispersal of seeds was the original, primitive technique and animal-aided dispersal a more recent development. But several researchers now believe that animal-aided dispersal is in fact more ancient. The cycad trees, survivors of the earliest group of seed-bearing plants, have large, heavy seeds poorly suited to travel on air. Such bulky seeds seem to have been common among the earliest seed-bearing plants that remain only as fossils. These early seed-producers lived in the tropics, where animal life abounded; presumably they found willing agents of dispersal there. The wind-dispersal techniques, under this theory, arose under the same conditions that promoted wind pollination, when seed-bearing plants spread into the cold northern habitats that supported little animal life on land. Even today, land plants in the far North, such as birch and willow, generally have tiny lightweight seeds that ride the wind rather than relying on chance encounters with animals.

Animal-assisted seed dispersal can happen in one of three ways. If the seeds are light enough and spiny enough, they may cling to animal coats and get a free ride on them; this stratagem is known as epizoochory (literally, "seeds on animals"). Alternatively, the seeds may be tempting enough tidbits to incite animals to collect them into a cache where some of the seeds will sprout—if the animal doesn't return to eat them, or fails to clean out the cache. This technique, which we might call the lazy squirrel syndrome, is termed synzoochory ("seeds with animals"). Last but certainly not least, the seeds may be covered by a soft fleshy coat that entices animals to eat them immediately. The fleshy coat is digested, but the precious seeds emerge unharmed, either spat out or voided in droppings. This last technique, a relatively risky one for the seed, is known as endozoochory or "seeds inside animals," but we might well call it the Jonah syndrome.

Each of these techniques demands a different adaptation from the animal partner. To get a free ride, the seeds must have spines, hooks, or sticky surfaces with which to latch on to their hosts. The animals, by the same token, must be fuzzy, furry, or feathery so that the seeds have something to cling to. Burdock cockleburs, covered with tiny hooked spines, are a beautiful example of free riders. Used by children for generations to make cups, bowls, and balls, the burrs cling to each other like Velcro. (In fact, burrs were said to have been the inspiration for Velcro.) They also stick tightly to human hair and clothes. In Beverly Cleary's *Ramona and Her Father*, one of the masterpieces of modern children's literature, the heroine crowns herself with a tiara of cockleburs; her father spends the rest of the day removing it, burr by burr, from her hair. Painful as such experiences are, they bear witness to the stick-to-it-iveness of burdock seeds.

The lazy squirrel syndrome occurs in a wide range of nut-bearing trees. Squirrels, chipmunks, and even certain birds—the acorn woodpecker in the western United States, for example—gather and cache nuts. But anyone watching a squirrel hunt for buried nuts in midwinter knows that many caches are never found again. A few nuts in these underground sites germinate, so that an oak seedling with a little help from a squirrel can sprout in the middle of a meadow.

Lodgepole pine and Douglas fir, discussed in Chapter 2 for their methods of defending their seeds against squirrels, also take advantage of the lazy squirrel syndrome when squirrels do succeed in gnawing through their cones. Another western tree that depends on seed-gatherers is the piñon pine. In this case, the major animal dispersers are birds, especially a large one known as Clark's nutcracker. Field study shows that an adult nutcracker can collect an average of fifty-five seeds in a single visit to a piñon tree. The seeds are stored in a special pouch beneath the bird's tongue—the nutcracker's version of chipmunk cheek-pouches—and then carried away to an underground storage site. These caches are found at distances of over twelve miles from the tree on which the nutcracker fed. In each cache, normally several seeds germinate and only one of these seedlings survives.

The piñon pine produces seeds of varying degrees of healthiness. Some, damaged by insects or fungus disease, are less likely to germinate than others. The damaged seeds are light tan in color, while healthy ones are dark brown. The birds discriminate between these colors and prefer the darker seeds. Perhaps the damaged seeds are less tasty or less nutritious. The piñons, in some way, have succeeded in reserving the birds' valuable labor for the seeds that are really likely to germinate.

The nutcracker performs an additional quality-control check using a peculiar behavior pattern called bill-clicking. After popping a seed into its beak, the bird shakes its head rapidly to produce a rattling noise in its mouth. They spit out the injured seeds, which apparently sound different from healthy ones, and pocket the healthy seeds.

The piñons have adapted the placement and shape of their cones too. The seed cones grow upright on branches where birds such as nutcrackers and jays can easily peck at them, instead of hanging down like the cones of wind-dispersed pines. Piñon cones have relatively thin seed-covering scales, easy to peck off. And in addition the cones open spontaneously but gradually. Each cone will release only a few seeds at a time on each tree, which supplies the birds with a prolonged source of seeds. In contrast, wind-dispersed pines open each cone's seed scales all at once, dumping the seeds out en masse onto the wind.

A special case of dispersal through seed caches occurs in seed-bearing plants that depend upon ants. It was known at the time the Book of Proverbs was written that some ants cache seeds. Hence the suggestion, "Go to the ant, thou sluggard, consider her ways and be wise; Which having no guide, overseer or ruler, Provideth her meat in the summer, and gathereth her food in the harvest" (Proverbs 6:6–8). Later naturalists doubted the accuracy of the writer, but as a twentieth-century ecologist has pointed out, these skeptics almost all lived in northern Europe, where ants never store seeds in large quantity. In Israel, Lebanon, and Syria, however, the countries that made up the old land of Canaan, there are ants that do. And careful field study has shown that ants in many other parts of the world also gather seeds—not in large enough quantities to attract the average person's attention, but large enough to ensure seed dispersal for the plants involved.

Plants that depend on ants for seed dispersal, technically termed myrmecochorous ("ant-carried") plants, produce seeds with features well-suited to their tiny partners. First and foremost, the seeds are small. They are also usually white or very light in color, easy for the near-sighted ants to detect against a dark seed pod or the ground. Most have a mass of oil tissue, known as the elaiosome or oil body, that attracts ants. The elaiosome contains a mixture of fatty foods and sometimes a little sugar or starch as well. One researcher suggests that the elaiosome may also contain fragrant or aromatic oils that help alert the ants' sense of smell to the seed. In the laboratory, ants usually reject seeds that have had the elaiosome cleaned off.

Examples of ant-dependent seed-bearing plants from the United States are violets, bloodroot, wild ginger, hepatica, and other small forest flowers. All of them set seeds early in spring, when ants are just coming above ground from their winter dormancy, eager for food. There is little other ant food available at that season, so that myrmecochorous plants get full attention. Their seed pods are close to the ground, accessible to the ants. In one case, the wild cyclamen, the flower faces upward to attract pollinators;

after seed-set, the flower stalk twists so that the seed pod bends down near the ground, ready to attract ants.

These small wildflowers, it turns out, reap more than one benefit from their ant partners. A major advantage is escape from other seed-eaters that would devour the seeds on the spot rather than carrying them to a cache as the ants do. In one study, rodents in a West Virginia forest ate up to seventy percent of the seeds from wild ginger and bloodroot when ants were not allowed access to the seeds. If the ants were permitted to gather and cache the seeds, the rodents destroyed only about thirty percent. White-footed deer mice, which are adept at locating and consuming small seeds, had a hard time finding the seeds concealed by the ants.

A different advantage is conferred by the cache site itself. The soil around an anthill, usually soft, holds more water than the average patch of forest floor, and a recent study found that the soil around an anthill is also richer in nitrogen. Both of these conditions encourage good growth for the seeds that germinate there.

Just how common is ant dispersal? In a recent field study, almost one-third of all the herbaceous plants in a West Virginia forest were shown to be myrmecochorous. And a botanist working in Australia estimates that as many as 1,500 species of plants there may be ant-dispersed. In the dry, nutrient-poor soil typical of many parts of Australia, the extra capacity for holding water and the extra nitrogen in anthill soil become crucial factors for seedling survival. Plants compete intensely on these ant mounds, with one species usually king of the hill.

In warm, moist tropical forests, ant dispersal takes a

peculiar form. Ants there gather seeds from flowering vines to form caches in their nests—but their nests, in many cases, are built high up on a tree-trunk. Made of chewed-up leaves and bark, the soft fibrous material of the nest is spongy and holds water very well, providing an excellent growing medium for the stored seeds. Thus an ant nest may become an ant garden, festooned with flowering vines that would have perished if the seeds had been left to struggle on the dark forest floor.

When a forest habitat originally well-adapted for ant-dispersal is disturbed by lumbering, the coevolved partnership breaks down. Field study showed that in such a disturbed northern forest, most of the bloodroot seedlings sprouted very near their parent plants. As a result, densely crowded stands of bloodroot formed, with all the disadvantages of easy discovery by enemies.

Under good conditions, an advantage of seed dispersal by ants is that the partnership is a generalized one. As several writers have pointed out, almost any myrmeco-chorous plant supplies the same food to a hungry seed-gathering ant. At the same time, almost any species of seed-gathering ant can glean among the seeds of an ant-dependent plant. This flexibility helps to promote the survival of the partnership.

The most extreme claim yet advanced for the benefits of ant-assisted seed dispersal was presented in a paper by the biologist Rolf Berg on the four species of wild barberry that grow in the Pacific Northwest. All of these barberries seem to have a common point of origin in the northern part of Asia, once connected to North America by a land bridge. Barberry seeds appear to be ant-dispersed. Berg hypothesized that their migration to the New World came about by ants moving the seeds across that Bering Strait land bridge!

The Jonah syndrome, or endozoochory, is a riskier procedure for seeds and spores than the first two dispersal strategies. It calls for the plant to produce an attractive fruit or fruiting body, ideally bright in color or tempting in smell as well as tasty. It also demands that the seeds be tough enough to survive a rough trip through the alimentary canal, and be ready to germinate when they hit the outside air again. But there are advantages that compensate for the risks. If the seeds or spores are voided in the animal's droppings, they gain some nitrogen as the droppings decay around them. In addition, many herbivores avoid eating plants that grow in dung, so the seedlings gain some protection from their surroundings. In cow pastures, the infiltration of weeds among the grass often begins with the sprouting of seeds from a pat of dung.

Until recently biologists believed that flowering plants and their seeds were the only participants in this method of dispersal. Certain fungi, especially the stinkhorn mushrooms, were known years ago to participate in the free ride syndrome. Stinkhorns attract flies to visit their foul-smelling fruiting body, and the flies carry the mushroom spores off to other spots. But we know now that an important group of underground-fruiting fungi, the truffles and false truffles, disperse their spores via consumption by animals, very like fruit-bearing plants.

Truffles and false truffles are the fruiting bodies of underground fungi that are important for the survival of most forest trees, especially pines, firs, and spruces. Filaments of these fungi grow on or in the roots of these trees, and assist them in extracting moisture and minerals from the soil. In some cases the fungi actually create fertilizing compounds for the tree roots. Such associations between fungi and roots are called mycorrhizae or "fungus-coated-roots." For most of the year the fungi live and grow as a thin webbing of pale threads among the roots, but when weather conditions are just right—warm and moist—the fungi rapidly produce fruiting bodies loaded with spores. Many of these push up through the soil to emerge as our familiar woodland puffballs and mushrooms. The wind disperses spores from these. But the truffles and false truffles form fruiting bodies that stay underground. It is easy enough to see how the above-ground or epigeal fungi travel far and wide—but how are the underground or hypogeal species dispersed?

The answer lies in the aroma and taste of underground fungi, and the terrific appeal they have as food for many wild animals. Pigs, and even dogs on occasion, have long been used to locate truffles in European forests. In this country, small forest rodents have a voracious appetite and a keen nose for our native truffles, although these appear to lack the appeal of European truffles to human taste buds. When two species of small rodents, the Oregon vole and the chickaree, were fed fruiting bodies of underground fungi in the laboratory, their droppings contained living spores that germinated. In fact the California red-backed vole appears to eat truffles exclusively. Analysis of the stomach contents of trapped specimens revealed almost nothing but truffles in their diet.

Voles and chickarees are ground-dwelling mammals, but studies of stomach contents showed that even tree squirrels, which feed heavily on seeds and bark, sometimes dig up underground fungi for food. So do flying squirrels, and this helps to explain a long-standing minor mystery in forest ecology: why are flying squirrels sometimes found in the gut of ground-dwelling predators? We can assume now that in addition to the occasional glide that turns into a fall, flying squirrels are vulnerable to capture when they are digging up truffles. Probably in the same frame of mind as a person about to enjoy an ice-cream sundae, they are not really on guard against their enemies.

The benefit to the forest trees and to the fungi of such mycophagy or truffle-eating is tremendous. Sterile soils are inoculated with beneficial fungi through the rodent droppings. The inoculated soils promote good growth in tree seedlings, since the fungi cling to newly arrived seeds and form helpful mycorrhizae around their roots. As James Trappe and Chris Maser, two of the outstanding researchers on this subject, point out, it is no longer appropriate to think of small rodents as forest pests. Although they destroy some seeds, their role as dispersal agents for mycorrhizal fungi more than makes up for this. The relationship is actually a three-way partnership. The rodents get nutrients from the fungi, and disperse some of their spores. The fungi help the trees grow, and the trees give the fungi sugars which come down from their leaves. The trees benefit from the rodents' fungus dispersal, and give them seeds as well as shelter in their trunks and branches.

The seed-bearing plants that run the risky course of dispersal by animal consumption are protected in a number of ways. First, the seeds themselves are tough-coated, and thus resistant to digestion in an animal's gut. Many such seeds are distasteful or actually poisonous, so that even a hungry animal would be reluctant to munch on the seeds for their nutrients after chewing up the fruit around them. Classic examples of such seeds are the pits of peaches and plums, and apple and pear seeds: raw, these all contain hydrogen-cyanide-releasing compounds! Cooking drives off the poison, so that applesauce made from unpeeled, uncored apples is harmless.

The pioneers among vertebrate animals in seed-dispersal by fruit-eating, it has been suggested, were the plant-eating dinosaurs. Huge, and with relatively small teeth, such creatures certainly could have consumed and dropped large numbers of seeds. And they coevolved with the flowering plants on land. The avocado, with its massive pit and oily flesh, is thought to be a modern survivor of plants that were once dispersed by big repiles or big primitive mammals like the ground sloth. But examples of reptile dispersal today are few and far between. Giant tortoises and iguanas on the Galapagos Islands in the south Atlantic eat fruit from the cacti there and disperse their seeds. Closer to home, the small North American tortoise may be the dispersal agent for wild strawberries. The berries are fragrant and grow near the ground, often masked by leaves above. Tortoises have an excellent sense of smell, like to eat soft fruit, and are low-slung herbivores, able to probe among the leaves.

In some special environments, fish have evolved into important dispersal agents for riverbank trees and shrubs. This seed-eating behavior is known from a number of river fish in the East Indies and in the Amazon river system in South America. Periodic flooding in the Amazonian jungle frequently carries fish over the banks, allowing them to swim far away through the forest, so that the seeds they drop are dispersed great distances. Some fish-dispersed seeds are tiny, like those of wild figs. Others are large, heavy, and brightly colored, attractive enough to be used as bait by local fishermen. These larger seeds are tough but have thin, edible coatings of fleshy tissue. Seed-eating fish chew off the coatings using peculiar wide, flat teeth reminiscent of the grinders of a cow.

Forestry operations conducted on a huge scale throughout Amazonia in recent years now threaten the riverbank partnership between trees and fish. In many areas the number of trees has been severely reduced, and this in turn reduces the food on which these seed-eating fish depend. Marked drops in freshwater fish populations have already been detected in the areas of heaviest deforestation.

The most successful vertebrate seed dispersers are seed-eating and fruit-eating birds. Winged, highly mobile, and in need of large quantities of food to stoke their metabolic furnaces, they are beautifully suited to the dispersal of seeds by endozoochory. In addition, they are toothless and thus not so likely to damage seeds in transit.

Most bird-dispersed seeds have a red or orange outer coat. This coat may be fleshy, as in apples, cherries, and raspberries, or it may be a thin layer of fatty tissue known as a funicle or aril. The large heavy brown seed of nutmeg trees falls into the second category: it is enveloped by an orange aril that attracts the birds. (Nutmeg arils, when harvested by human hands, are dried and ground to yield the spice called mace.) Red and orange fruit, highly visible against green foliage, attracts birds, which have excellent color perception. A few plants have exploited this fact by evolving large, tough seeds with no edible outer coating at all, which are streaked with red or black. Such seeds, ecologists believe, attract birds by their deceitful resemblance to the seeds with nutritious arils.

Bird-dispersed fruits generally have no fragrance; this reflects the birds' poor sense of smell. In temperate zones like most of North America, where fruit-eating birds are small—such as cedar waxwings and mockingbirds—the seeds of their coevolved fruits are also small, such as raspberry and blackberry. In the tropics, where larger fruit-eating birds are common, many fruits that depend on birds have larger, heavier seeds, such as loquats and kumquats.

Although most bird-dispersed seeds have very tough seed coats that enable them to resist digestive juices, there are a few exceptions. One of the most striking of these is the mistletoe family. The seeds of many of these parasitic plants are soft, but they are coated with a bitter, sticky material called viscin, as well as a layer of soft flesh. Birds that feed on mistletoe eat the berries, swallow the flesh, and regurgitate the viscin-coated seeds. The viscin allows the seeds to cling to a branch of the tree where the bird is perched; later they germinate on the branch. A number of birds are specially adapted for mistletoe-eating by having no gizzard at all in their gut, so that even if they swallow the delicate, distasteful mistletoe seeds, these usually survive and emerge unharmed in the droppings. The bird-dependent habit of mistletoe seems to have been recognized in Europe hundreds of years ago: the Old English word "mistletoe" means "thrushes' droppings."

In cases where a seed-eating bird does have a tough gizzard, a plant may coevolve seeds so rugged that they have difficulty germinating without being processed through the bird's gut. In such cases a trip through the gut scrapes away some of the excess coating, permitting moisture to enter the seed, and promoting germination after the seed hits the soil. A South African shrub, *Acacia cyclops,* illustrates this adaptation. Introduced to South Africa from Australia in the 1840s, the shrub now covers thousands of acres. Its seeds are dark brown and tough, with a reddish funicle. The seeds from this plant that are dispersed in bird droppings germinate nearly twice as well as seeds merely picked from the shrubs and planted by hand. The droppings, in the poor, sandy soil of South Africa, also help promote *Acacia* growth by supplying extra nitrogen to the seedlings.

There are plants that have evolved such an intense dependence on their bird partners for seed dispersal that their seeds cannot germinate at all unless scarified by the trip through the bird's gut. Such plants are skating on very thin ice, so to speak. If their bird partner vanishes, their seeds will fail to sprout even when they hit the soil. Onc such case is known from the island of Mauritius in the Indian Ocean, the home of the now-extinct dodo. Human hunting eradicated the dodo a hundred years ago. The discovery has since been made that a large tree of Mauritian forests, *Calvaria,* has failed to reproduce itself for nearly as long. *Calvaria* trees do set seed, but their large, heavy seeds lie ungerminated in the soil. This seems to be a direct consequence of the extinction of the dodo. The large, heavy beaks and strong muscular gizzards of the dodos would have abraded the seeds and readied them for germination. Without this aid, each seed embryo now is doomed to death by starvation in solitary confinement.

In experiments designed to test this hypothesis, *Calvaria* seeds were fed to turkeys, the closest modern equivalent of the dodo in size, strength, and feeding habits (and perhaps intellectual ability). Ten out of seventeen pits fed emerged considerably scraped up, and three of these ten germinated when planted—the first *Calvaria* seedlings in living memory. Tolkien fans may see a resemblance in this real-life drama to the story of the Ents, who had lost their Ent-wives and therefore had had no Entings for "a terrible long time" —but the *Calvaria* story may end, or at least go on, happily. Hand-scraping or turkey-gizzard abrasion of the seeds can help keep *Calvaria* from extinction.

Many birds, of course, visit seed-bearing plants and feed on them, but not all are necessarily true seed dispersers. A field study in Costa Rica clarified this point by identifying the birds that visited one species of tree, *Casearia corymbosa*, and recording their behavior while on it or nearby. The goal was to discover which of the many bird visitors were the real dispersal agents for this tree.

Analysis of the data revealed that only one of all the twenty-two species of birds recorded from *Casearia corymbosa* is an effective partner in its seed dispersal. This bird, known as the masked tityra, took more fruits per visit than any other species. It normally flew to trees up to 200 meters away before swallowing the fruits. Thus the seeds, which were then regurgitated, fell to the ground a good distance from the parent trees. Other visitors ate their fruit and regurgitated the seeds while perched on the parent trees. Their seeds fell on the ground beneath the parent tree, and, as might have been expected, did poorly. In addition, other visitors were only sporadic consumers of *Casearia* fruit, but the masked tityra visited these trees consistently throughout its fruiting season. This species, then, can be identified as the only real seed dispersal agent for *Casearia corymbosa.* The other species, in the researcher's words, are merely "fruit thieves."

Dispersal of seeds by mammals that eat the fruit, while less common than endozoochory by birds, does happen, especially in the tropics. In the East Indies large carnivorous mammals eat and disperse the seeds of the large, awful-smelling durian fruits. And in South America, jaguars and other carnivores eat the oily, high-protein fruits of avocadoes and disperse the heavy seeds. But the outstanding mammalian partners in the Jonah syndrome are monkeys, apes, and bats.

Monkeys and great apes are fond of fruit, dexterous, and highly mobile in their daily search for food. Armed with strong teeth and in some species able to climb far out on slender branches to reach their foods, they are excellent candidates for seed-dispersal partners. The hamadryas baboon, a powerful ground-dwelling monkey native to northern Africa, acts as a dispersal agent for prickly pear cacti there. Sizable hedges of prickly pear cactus surround one of their resting sites in the cliffs of northern Ethiopia. The hedges grow from prickly pear seeds, voided in the baboons' droppings after they feed on the cacti some distance away. Howler monkeys in South America and cebus monkeys in Africa have also been implicated as dispersal agents for fruiting plants. The fruits they and the great apes favor are generally colorful—red, orange, or yellow—and often fragrant. Both qualities reflect the animals' sensibilities: monkeys and great apes have excellent color vision and a good sense of smell.

The fact that monkeys and apes love fruit is no surprise to anyone who has seen Jane Goodall's films of chimpanzees frantically gathering every banana she offered them, like hungry wedding guests filling their dinner plates at a long-overdue reception. But most of us in the temperate zones find it hard to picture bats eating fruit. The small, insect-eating bats of North America are indeed poorly suited for seed dispersal. But the large, fruit-eating bats common in tropical Africa and in Australia are a different story altogether. Anyone who raises fruit-bearing trees in these regions is likely to confront a flock of voracious fruit-eating bats once the crop ripens, and may have to watch in fury while they devour the ripe mangos, guavas, or figs he or she was planning to harvest the next day.

Fruits dependent on these bats for seed dispersal are usually soft-fleshed and drab or green in color when ripe, adaptations evolved to suit the bats' small weak teeth and poor color vision. Fruit-eating bats appear to locate their fruit mainly by its smell. Most of the bat-dispersed fruit gives off a strong sweet odor, or else the same musty, "batty" smell that attracts flower-feeding bats to their coevolved flowers.

Fruit-eating bats are large and strong, bigger than many fruit-eating birds. One group of such bats is known as flying foxes, and they look the part. The seeds they disperse are frequently larger and heavier than bird-dispersed seeds. The trees offer fruits to the bats in a manner reminiscent of flagelliflory among bat-pollinated flowers: the fruits hang on the ends of branches or dangle on long stalks below them. This positioning makes it easier for the bats to get at the fruit without tangling their wings in the branches.

Fruit-eating bats often nest communally; one field study

recorded up to two hundred bats sharing a roost each night. The bats foraged mainly on fruit-bearing trees near the roost, thus saving on energy expended in commuting.

In another study of foraging behavior, bunches of several kinds of fruit were tied to long poles placed throughout a tropical forest. For certain fruits, the likelihood of being gathered by bats was greatly increased when the fruit-pole was placed near a natural stand of the same kind of tree. But for other species the reverse was true: bats found the fruit-pole more often when it was *not* near a natural stand. These differences grow out of the bats' perception of differences in distribution of the trees involved. Some fruit-bearing trees are very patchily distributed in groves; others are more uniformly spread out. Bats feeding on patchily distributed groves tend to divide their time into bouts of feeding and bouts of commuting. But those that feed on evenly dispersed trees fly a little and eat a little, off and on. This pattern of foraging would be more likely to lead bats to find poles near groves of the same kind of tree.

Some fruit-eating bats are active by day, but many, like bats in general, are active only at night. One researcher working in the tropical forests of Panama discovered that forays by fruit-eating bats were most frequent on nights during the dark of the moon. Very few of these bats are on the wing on bright moonlit nights. This seems rather curious at first glance, since the light would help bats find their fruit. The ecologist who reported this lunar phobia, as he called it, explained the behavior as a way for the bats to escape predation. Their "fear of flying" when the moon shines brightly keeps them out of sight of nocturnal predators, such as owls, that hunt by sight.

Another field study in Panama used radiotelemetry to

measure the effectiveness of fruit-eating bats as seed-dispersal agents. Small signal-emitting radios were attached to the fur of twenty-two wild bats, and monitored during their nighttime flights. The results showed that the bats first flew to a fruiting tree, and then moved to perch on other trees, before they ate their harvest. The bats showed a marked preference for figs: about eighty percent of their droppings contained fig seeds. The fig, in a co-evolved technique of offering constant reward to its bat visitors, produced ripe fruit throughout the year. Howler monkeys in the same forest also ate figs, but they consumed the fruit when it was green; the bats selected only ripe figs containing mature seeds that germinated well. Thus the bats are effective dispersal partners for the fig, while the howlers could be labeled, as in the *Casearia* study, fruit thieves.

A follow-up study showed that the fig-eating bats frequently use a peculiar method of feeding termed the "juice-only" technique. In this method, the bat chews up a fig, swallows its juice, and then spits out a compressed mass of seeds and fiber which makes up most of the fig's dry weight. The swallowed juice, rich in minerals and dissolved sugars, is very low in protein, but these bats appear to eat a number of insects as well as fruit, which supply the fat and protein needed to balance their diet. By using the juice-only technique the bats are able to process twice their body weight in fruit per night.

Seed-dispersal partners mentioned only briefly in most surveys of the subject, but of great economic importance, are human beings. As with pollination, many cultivars—corn, tomatoes, roses, and grapes, among others—owe their continued existence to human propagators. In fact, many cultivars are now exclusively reproduced by cuttings, not seeds. In the process they have evolved tiny, useless seeds, like those seen in "seedless" grapes and bananas. Unfortunately for the future of these plants, not only is there just one species on earth—our own—still able to keep them going, but most humans in modern societies lack the horticultural skill even to do this. Thus the survival of human-dependent plants rests in the hands of a small cultural elite, rather than—as with ants, birds, and bats—depending on an entire population of willing partners.

CHAPTER 10

Whither Thou Goest: Long-Term Animal–Plant Symbioses

Whither thou goest I shall go,
and where thou lodgest I shall lodge;
thy people shall be my people,
and thy God my God.
RUTH 1:16

THE ANIMAL–PLANT MUTUALISMS SEEN IN POLLINA-
tion and seed dispersal involve only brief contact between
the partners. Seeds stay with their dispersal agents only a
few days at most; pollen stays with its disperser only hours
or minutes. And this is as it should be: seeds and pollen get
no younger in transit, and aged seeds or pollen grains are
less lively than young ones. But in some mutualisms the
association is longer—it is, in fact, for life. Termites and
ants that grow fungi for food never abandon their crops;
even when their queens are on the wing, pellets of fungi
go with them. Reef-building corals, giant clams, and other
marine animals harbor tiny algae that actually live beneath
their host's skin. Such mutualisms, once established, are

immensely important for the survival and growth of the animal host, especially if its environment is nutritionally poor. The symbionts help to feed their host.

On dry land, the best-documented of such symbioses, or mutualisms, are the ones that occur in wood-eating insects. Lignin and cellulose, the main ingredients in wood, are very tough to digest, and none of the wood-eating insects secrete enough digestive juices to turn the wood into sugar. But certain protozoa and certain fungi can. The wood-wasp is one well-studied insect that depends on such aid. A mated female wood-wasp drills into a tree trunk and lays her eggs in the tunnel. As the eggs slide down they pick up tiny fungal cells, stored in pouches at the base of her reproductive system. The fungi grow quickly on the wooden walls of the egg-storage tunnel, and when the larval wood-wasps hatch out they eat the fungus-softened wood. (If separated from their fungal partners and fed plain wood, they starve.) The growing larvae catch fungal cells in the folds of their soft skin. Later, the metamorphosis to adult form includes infolding of such creases to create fungus-holding pockets at the base of the female reproductive tract—and the cycle goes on.

Wood-eating beetles known as ambrosia beetles have evolved a different strategy that also involves a fungal partner. The ambrosia beetles carry their fungi around in tiny pouches—mycetangia, or fungus-pockets—on their body surface. After chewing into a tree trunk, the beetles rub their fungi onto the tunnel walls. In short order a dense growth of fungus sprouts, and the beetles, young and old, eat this growth which, though it hardly suggests the food of the gods, has long been termed ambrosia.

If the ambrosia beetles are separated from their fungi,

both partners suffer. Juvenile beetles do mature, but the adult females are sterile. More than one specialist has suggested that ambrosia not only digests the wood but also secretes a vitamin necessary for fertility in its insect partner. And the abandoned ambrosia is rapidly overrun by competing species of fungi, just as weeds invade the lawn of a deserted house.

The special fungus-pockets, or mycetangia, on ambrosia beetles seem to have evolved from pouches that once held oily fluid secreted as a lubricant for tunneling. Even now, the mycetangia still secrete oily fluid; this helps keep down the growth of weed fungi when it is smeared on the tunnel walls.

Fungus-raising termites have a far more elaborate social organization than the wood-wasps or ambrosia beetles, but they have just as deep-seated a reliance on fungal partners. The family of termites that grow fungi, the Macrotermitinae, surpass all other insects in their ability to alter their environment, since they build huge clay nests—castles of clay, one researcher calls them—that can rise ten feet or more above soil level. The inner structure of these clay castles is beautifully suited for growing fungi. In the dark, warm, moist passageways, the termites raise their fungi on beds of chewed wood and leaves, manured with their own droppings.

The termites' fungi have become every bit as dependent on their care as ambrosia fungi are on their beetles. If a termite nest is destroyed or the insects in it are poisoned, competing species of fungi soon overrun the cultivated one, which dies off.

Is the same intense dependence true for the termite partners? Some researchers report that termite colonies can

live in the laboratory with only sterile "comb" for food—
the material on which fungi are grown—and no live fungi.
Other researchers, attempting the same experiment, have
seen their termite colonies dwindle and die out. Recent
analysis of the digestive juices found in at least one macro-
termite species has revealed that a crucial digestive juice is
largely secreted by their fungus. When the termites were
deprived of contact with the fungus this juice, essential for
digesting wood, fell to only fourteen percent of its normal
level. But relative dependence of termites upon such external aids to digestion may vary from species to species,
even from colony to colony. That would account for the
differing results of deprival experiments.

Leaf-cutter ants, known as attine ants, are even more
sophisticated fungus gardeners. They too grow fungi on
plant material within their nests, but while the termites
use chewed-up wood and old leaves, the attines use
chewed-up insect droppings and pieces of fresh leaves to
grow their fungi on. Attine ants are found only in the
New World; they are commonest in the tropics, although
some species live further north, in New Jersey.

The behavior that makes them a pest of man was first
analyzed in 1874 by the English naturalist Thomas Belt.
The hypothesis he proposed he acknowledged was "extraor-
dinary and unexpected": the cut-up leaves are used in the
nest to grow fungi on which the ants feed. In support of
this idea, he dug up several large nests and found that
innumerable chambers were filled with soft spongy material. It was composed of chewed-up leaves, overgrown
with tiny fungi.

Later research has confirmed that the attine ants do indeed feed on their fungi. Under ant cultivation, the fungi

produce tiny swellings, variously known as gongylidia, bromatia, or kohlrabi bodies. The ants nip off these little lumps and eat them.

Dispersal of the fungi is carried on by female ants. When a virgin queen leaves the colony on her mating flight, she carries with her a dowry of fungal spores. These are pressed into a tiny pouch at the back of her mouth, which thus functions as the mycetangium does in ambrosia beetles. After she mates on the wing, the young queen descends to earth, digs a burrow, spits out the spore packet into it, and thereby starts a small fungus garden. She lays eggs in the burrow, and the offspring that hatch from these eggs help her care for the garden. Later on they take over this chore entirely, while the queen devotes herself to laying eggs.

Care of the garden involves fertilization and pruning of the fungi and weeding out of undesirable species. The attine ants, in addition to providing chewed-up leaves for their fungi to grow on, fertilize the fungi with liquid droppings. In at least one casc, these droppings are essential for fungi unable to extract nitrogen from leaf material. The ants' droppings are rich in enzymes that break down leaves and in simple nitrogen-containing compounds as well. The partnership here is truly a two-way street; the ants help their fungi to survive, and the fungi feed their ants.

The selection of leaves on which the ants grow their fungi is a more complex process than it might seem. In one study, the ants turned out to use less than a third of all the plants available in their vicinity. And the commonest plants were seldom the ones most frequently selected. In fact, two of the eleven colonies studied showed a distinct tendency to use the rarer plants most often. Chemical composition of the leaves, notably water and mineral content, appear to influence their choice of material.

The ants' dependence on their fungi parallels that of human cultures on their crops. The fungi involved have been selected for high productivity but not for toughness, so that they fail in competition with rivals from the wild if their owners abandon them. And the ants involved have become so dependent on their fungi that their motto could well be "Fungus is the staff of life."

An even more extreme degree of symbiotic involvement is seen in the insects that keep fungal symbionts inside their bodies. Many sapsucking and bloodsucking insects, and some of the wood-borers, harbor such hidden partners. These fungi live in special cells called mycetocytes, or fungus-holding cells: they appear to create vitamins for their hosts. In the case of the drugstore beetle, so called because it is found in tobacco and stored drugs, if their endosymbionts were removed, the beetles grew poorly and had difficulty in moulting. Few of the experimentally deprived larvae reached adulthood.

A case that balances on the fence between mutualism and antagonism is the relationship of a tiny scale insect with its companion fungus. This fungus grows in soft velvety masses on the bark of trees. Scale insects burrow into it, and gain shelter from their enemies. But the fungus sends out slender threads or haustoria that plunge into the scale insects' bodies and suck nutrients from their blood. The individuals that are attacked survive, but fail to reach sexual maturity. Luckily, in every fungus, a few scale insects escape this victimization. They reproduce and their offspring fly away. But spores of the fungus ride on the escapees, and thus establish new colonies elsewhere.

Many aquatic animals harbor internal symbionts that are clearly beneficial: tiny green algae, related to the free-living forms that can make a pond look like pea soup. The grass-green algae may be so abundant that the whole animal is colored deep green. These symbiotic algae show some peculiar adaptations to their sheltered existence: unlike their own nearest free-living relatives, they have no eyespots or means of locomotion. When raised apart from their hosts they do form these structures. They also secrete protective cell walls, which are absent when they live as symbionts.

Even more common as algal symbionts are golden-yellow or brownish algae that belong to a group called the dinoflagellates. These algae inhabit giant clams, reef-building corals, and various jellyfish, worms, and other marine animals. They too can be present in overwhelming abundance, enough to color the animal golden yellow or chocolate brown. Most writers on marine symbioses refer to this group of algae as zooxanthellae, "little yellow things in animals." Despite their color, these algae too contain chlorophyll, and like their greener relatives they release significant quantities of sugar and oxygen if illuminated.

The sugar released by algal endosymbionts can mean the difference between life and death to their host. If the little freshwater polyp *Chlorhydra* is kept under lights in the laboratory, it survives much longer with the green symbionts it harbors than when these have been removed. One study suggested that the *Chlorhydra* symbionts raise their output of sugar when the host is starving.

Giant clams are normally found in Pacific Ocean waters

that are low in available food. Yet they are notorious for growing far bigger than their nearest relatives, the little cockles. Their record sizes—the shells in one species grow to a length of four feet, and have been used as baptismal fonts—may depend in part on their endosymbionts' active output of sugar. Studies in which the clams were bathed with radioactive carbon dioxide gas revealed that the algae converted the gas into sugar, and that the radioactive sugar turned up in significant quantities in the clams' tissues.

The giant clams keep their symbiotic algae in their blood vessels; one investigator thought the symbionts actually lived inside their blood cells. Aging symbionts, he believed, were digested by the clams. But recent studies found no symbionts inside blood cells. Although the aging algae do enter the clams' guts, they emerge in the droppings unharmed, still quite healthy. A point on which all hands agree is that the algal symbionts are valuable partners for the giant clams. On a bright sunny day the algae may contribute over eighty percent of all the sugar the clam needs. Even in cloudy weather the algae are able to synthesize enough sugar to meet over sixty percent of the clam's metabolic needs.

The thin layer of skin that secretes the shell of clams, the mantle, is particularly rich in algal-holding vessels. The thick shell over it, of course, excludes light. But the giant clams get around this blackout by a behavioral adaptation. Unlike most other clams, they position themselves with the shell's thick hinge downward, and the open edges of their shells up. They allow the soft thin edges of the mantle to extend far out into the water beyond the edges of the shells. They keep this position, even when the waters above grow rough, thanks to the heavy mass of the shells. The body plan of a giant clam is like that of an old-fashioned Russian wooden doll with a lead weight at the bottom. Its heavy hinge allows it to roll back into the same position no matter how hard it is rocked by the waves.

The transfer of symbiotic algae to young giant clams involves active acquisition by the free-swimming clam larvae. In one study, clam larvae picked up algal symbionts by swallowing them. The larvae were then maintained in well-lit seawater tanks for ten months, with no food except what their algae supplied. The young clams continued to grow and thrive on this watery diet.

Coral polyps, like the giant clams, show an intense dependence on symbiotic algae. Even though each polyp is a predator, the tiny creatures it stings to death supply no more than twenty percent of its energy needs. Symbiotic algae provide the remaining eighty percent, as with the giant clams. The algae play an essential role in coral reef formation, accomplished through the polyps' secretion of calcium compounds. Peak rates of calcification occur when the polyps are brightly lit, allowing the algae to photosynthesize at a high rate and provide the polyps with nutrients.

In addition to sharing sugar with their host, certain algae may release "growth factors" to them. Studies on a saltwater worm, *Convoluta*, showed that even when the subjects were all well-fed, worms with endosymbiotic algae grew faster than worms deprived of them.

Casseiopeia, a jellyfish, is another marine animal that plays host to algal symbionts. This jellyfish normally produces free-swimming larvae that settle, attach to rocks or seaweed, grow into plant-like polyps, and finally undergo a splitting-up into many thin layers or strobilae. Each strobila then floats back up to the surface to fatten into an adult jellyfish. If *Casseiopeia* larvae are deprived of their algae, they will not strobilate at the same temperature as normal larvae equipped with symbionts. The water has to be considerably warmer around them before they consent to split up. Thus the algae contribute something to *Casseiopeia* that promotes their strobilation—perhaps extra nutrients, perhaps a special growth factor.

Casseiopeia adults, too, have evolved behavioral adaptations that assist their algal partners. The algae are mainly concentrated in the soft underside of the body, and these jellyfish frequently float themselves upside-down in shallow water, thereby sunning their symbionts.

A tiny single-celled animal, *Paramecium bursaria*, plays host to even smaller symbiotic green algae; as if in recognition of their symbionts' need for sunlight, these protozoans regularly accumulate in well-lit areas. This behavior, known as photoaccumulation, vanishes in *Paramecium bursaria* deprived of their tiny partners. It is lessened but not destroyed if the algal populations are reduced but not totally eradicated.

Sea anemones that harbor algal symbionts also show finely tuned responses to light. Certain anemones move away from intense light, which damages the algal chlorophyll, when their normal symbionts are living in them. If deprived of the algae, they become indifferent to the level of lighting. Other anemones, which hide their body stalks in holes, harbor their zooxanthellae in long projections called pseudotentacles. (Their true tentacles, used to kill prey, are nearly empty of symbiotic algae.) During daylight hours, the anemones extend their pseudotentacles out of their holes, sunning their algae, while they keep the true tentacles pulled in. At night, the situation reverses: they pull in the algae-laden pseudotentacles and extend the true tenacles to catch what prey they can. Larvae of the closely-related coral polyps behave rather like the photo-accumulating *Paramecium*, congregating in well-lit seawater if their algae are present, and indifferent to lighting if the algae are absent.

What do the algae get out of this relationship? Shelter and sunlight, of course, but does this do anything for them that they would not get on their own? Studies on symbiotic algae isolated from their hosts in the laboratory suggest that the partnership does benefit the algae. Host-free algae continued to photosynthesize, but not at the rate they did while in their host. When bits of ground-up host flesh were dropped into the seawater around the algae, their photosynthetic output rose appreciably.

The specificity of the associations between symbiotic algae and animals has long been a hotly debated question. Some investigators are convinced that the symbiotic algae of most marine animals are merely varieties of one dino-flagellate species. Others, who acknowledge the great structural similarities of many of the symbiotic algae, are certain that chemical differences make only certain species of algae acceptable to each species of host.

An experiment designed to test this point used two species of the little freshwater polyp *Chlorhydra*. The species, *C. hadleyi* and *C. viridissima*, harbor nearly identical-looking green symbionts. But when deprived of their algae, polyps accepted only those replacements taken from polyps of their own species. More striking still, if a bleached *C. hadleyi* poly was grafted onto a green polyp of *C. viridissima,* the flesh of both polyps made a good union but no algae moved into the pale graft from the green stock. To each its own!

Another freshwater polyp, *Hydra viridis,* also harbors symbiotic green algae. In an attempt to find out how specific this association was, researchers deprived individual polyps of their symbionts, and then force-fed them algae from several different sources. Most were quickly vomited by the dissatisfied polyps. Out of thirteen kinds of algae, the only ones that were kept were those from other *Hydra viridis* polyps—plus, oddly enough, a group from *Paramecium bursaria.*

A later experiment along the same line tried to exchange algal symbionts between polyps of *Hydra viridis* and polyps of another species, *Hydra attenuata.* In this case, algae from the other species were sometimes initially accepted. But in the long run only algae from individuals of the same species were maintained as permanent symbionts.

One further experiment along this line involved a sea anemone, *Aiptasia pulchella* with symbiotic algae that look like those in reef-building corals and giant clams. When the anemones were bleached of their own algae and offered others, they accepted almost any algae they could get. But anemones that received algae from other *Aiptasia pulchella* showed far better growth, after reestablishing the partnership, than anemones that got algae from other hosts. In the second group—those receiving what the researchers called "heterologous" algae—there was an interesting twist. Anemones supplied with algae from giant clams did very poorly, scarcely any better than anemones left totally bleached. But anemones that received algae from jellyfish or corals fared a little better. Anemones, jellyfish, and corals are all in the same major animal phylum, while clams are far-removed in classification. The results here suggest that there is some acceptability of algal symbionts within a given animal phylum.

Establishment of the initial union between host and algal symbiont seems to involve contact between the cell surfaces of the partners. In a recent experiment, investigators treated *Hydra* polyps with the drug colchicine; they found that such drugged polyps failed to pull symbiotic algae into the proper position inside their cells. Colchicine, derived from the autumn crocus plant, has long been known to interfere with living cells' ability to make microtubules, tiny strands that promote folding-in and movement of cell surfaces. The algae thus may need to be actually tucked into place by microtubules in the *Hydra* cells.

In at least one case, the animal host appears to play no part at all in acquisition of its symbionts—the symbionts recognize the host, and move in. This seems to be what happens in the marine worm *Convoluta*: its free-swimming larvae naturally attract the appropriate algae, which settle on them and penetrate their skins. The worms' reward for accepting such invasion by algae is enormous. They never need to eat again, once the union is established. Many generations of *Convoluta* have been maintained devoid of food in well-lighted seawater in the laboratory.

The evolutionary origins of these long-term symbioses have still not been settled. Some biologists support the idea of origin through accidental association of plant and animal partner. Others believe that the relationships began as parasitisms that gradually turned into mutualisms. Still others suggest an origin by dietary intake.

The idea of an accidental association beginning a partnership has been particularly promoted for wood-boring insects and wood-digesting fungi. Air currents can accidentally introduce almost any fungal spore into freshly drilled tunnels in a tree trunk, or onto the newly fallen leaves on a forest floor. The fungi that grew best in these places would produce sugars, proteins, and vitamins. Insects that supplemented their ordinary leafy or woody diet with some of these fungi would benefit nutritionally; fungi that could hitch a ride on such insects would benefit by dispersal and by ready access to fresh plant fodder.

Another theory of symbiotic origins, proposed by the microbiologist Lynn Margulis, traces mutualism to parasitic beginnings. Her most famous application of this theory concerns mitochondria, the rod-shaped structures that release energy for higher plant and animal cells. Many biologists have assumed that mitochondria originated as specialized in-pouchings from the cell surface. Margulis believes that they began as tiny independent organisms, which at some point in the distant past invaded larger cells. At first purely exploitative, they later balanced the score by sharing energy-rich secretions with their hosts, and ultimately established a permanent symbiosis with them.

In support of this theory is the fact that mitochondria contain tiny ribosomes, cellular workshops that would allow them to secrete their own proteins. They also contain small amounts of DNA, the genetic material that provides the blueprint for these proteins. And their DNA structure differs markedly from that in the cell nucleus, their hypothesized "host's" DNA. Many biologists have accepted Margulis' hypothesis of mitochondrial origin, though some remain unconvinced.

Parasitism that turns into mutualism seems a less likely origin, though, for such endosymbioses as those seen in the marine animals and their algae. A third theory, that of dietary origin for the partnerships, seems more reasonable here. When marine animals eat algae, most of the algae are digested, but not always all of them. If a few eaten algae survived and later came to rest under the thin skin of their consumers, they would have the possibility of continuing to photosynthesize. The host would reap an unexpected harvest of sugars, oxygen, proteins, and vitamins. Algae tough enough to survive in the host would benefit by gaining shelter, dispersal, and body wastes from the animal to serve as raw material for their own protein synthesis—plus extra carbon dioxide to use in photosynthesis. In the long run, any adaptations or behavioral modifications that promoted the survival of each participant in this venture would promote the well-being of both.

The question of past origins remains an open one. What is the future of long-term symbioses? From the animal's viewpoint, the advantages gained are immense—energy-rich food from indigestible wood or sterile seawater. From the plant's viewpoint, the advantages gained seem less significant—shelter, transportation, nitrogen-rich wastes. The first impression is that—as in the fairy tale that gave this book its title—the Beast needs its partner more than Beauty needs the Beast. But little points we noticed in passing—the way algae are attracted to *Convoluta* larvae, the dependence of ambrosia fungi on their beetles' cultivation, the higher rates of photosynthesis in algae stimulated with host tissue—are all arrows flying toward the same destination. That destination is one of mutual dependence. In the fairy tale, Beauty came to realize that she needed the Beast as much as he needed her. In the biology of symbiosis, it seems likely that the plants involved will ultimately need their beasts just as much as the beasts need them, and that their long-term partnership will become for both the central fact of life.

Further Reading

The books, articles and research reports are arranged in the order in which they contributed to each chapter.

CHAPTER 1
Mutual Injury, Mutual Aid: Patterns in Coevolution

GILBERT, LAWRENCE E., and PETER H. RAVEN, eds. *Coevolution of Animals and Plants.* Austin, Texas: University of Texas Press, 1973.

OPIE, IONA and PETER, eds. *The Classic Fairy Tales.* New York: Oxford University Press, 1980.

MEEUSE, BASTIAAN J.D. *The Story of Pollination.* New York: Ronald Press, 1961.

CHAPTER 2
Self-Defense: Barbed Wire and Bodyguards

EHRLICH, PAUL R., and PETER H. RAVEN. Butterflies and Plants. *Scientific American* 216 (1967): 104–13.

GRANT, SUSAN. *Learning and Perception in Lepidopteran Larvae.* Unpublished doctoral dissertation. Amherst: University of Massachusetts, 1978.

PILLEMER, ERIC A., and WARD M. TINGEY. Hooked Trichomes: A Physical Plant Barrier to a Major Agricultural Pest. *Science* 193 (1976): 482–84.

GILBERT, LAWRENCE E. Butterfly–Plant Coevolution: Has *Passiflora adenopoda* Won the Selectional Race with Heliconiine Butterflies? *Science* 172 (1971): 585–86.

———. The Coevolution of a Butterfly and a Vine. *Scientific American* 247 (1982): 110–21.

RATHCKE, BEVERLEY J., and ROBERT W. POOLE. Coevolutionary Race Continues: Butterfly Larval Adaptations to Plant Trichomes. *Science* 187 (1974): 175–76.

ROTHSCHILD, MIRIAM, and L.M. SCHOONHOVEN. Assessment of Egg Load by *Pieris brassicae* (Lepidoptera: Pieridae). *Nature* 266 (1977): 352–55.

RAUSCHER, MARK D. Egg Recognition: Its Advantage to a Butterfly. *Animal Behavior* 27 (1979): 1034–40.

PROKOPY, RONALD J. Evidence for a Marking Pheromone Deterring Repeated Oviposition in Apple Maggot Flies. *Environmental Entomology* 1 (1972): 326–32.

SMITH, CHRISTOPHER C. The Coevolution of Pine Squirrels (*Tamiasciurus*) and Conifers. *Ecological Monographs* 40 (1970): 349–71.

BELT, THOMAS. *The Naturalist in Nicaragua.* New York: E.P. Dutton, 1911.

JANZEN, DANIEL. Coevolution of Mutualism between Ants and Acacias in Central America. *Evolution* 20 (1966): 249–75.

————. *Pseudomyrmex nigropilosa*: a Parasite of a Mutualism. *Science* 188 (1975): 936–37.

RICKSON, FRED R. Glycogen Plastids in Müllerian Body Cells of *Cecropia peltata* —A Higher Green Plant. *Science* 173 (1971): 344–47.

————. Progressive Loss of Ant-Related Traits of *Cecropia peltata* in Selected Caribbean Islands. *American Journal of Botany* 64 (1977): 585–92.

————. Developmental Anatomy and Ultrastructure of the Ant-Food Bodies (Beccarian Bodies) of *Macaranga triloba* and *M. hypoleuca* (Euphorbiaceae). *American Journal of Botany* 67 (1980): 285–92.

HUXLEY, CAMILLA. Symbiosis Between Ants and Epiphytes. *Biological Reviews* 55 (1980): 321–40.

BECKMAN, ROBERT L., JR., and JON M. STUCKEY. Extrafloral Nectaries and Plant Guarding in *Ipomoea pandurata* (L.) G.F.W. Mey. (Convolvulaceae). *American Journal of Botany* 68 (1981): 72–79.

BENTLEY, BARBARA L. Plants Bearing Extrafloral Nectaries and the Associated Plant Community: Interhabitat Differences in the Reduction of Herbivore Damage. *Ecology* 57 (1976): 815–20.

———. Extrafloral Nectaries and Protection by Pugnacious Bodyguards. *Annual Review of Ecology and Systematics* 8 (1977): 407–27.

SCHEMSKE, DOUGLAS W. The Evolutionary Significance of Extrafloral Nectar Production by *Costus woodsonii* (Zingiberaceae): An Experimental Analysis of Ant Protection. *Journal of Ecology* 68 (1980): 959–67.

Keeler, Kathleen. The Extrafloral Nectaries of *Ipomoea leptophylla* (Convolvulaceae). *American Journal of Botany* 67 (1980): 216–22.

RICKSON, FRED R. Absorption of Animal Tissue Breakdown Products into a Plant Stem—The Feeding of a Plant by Ants.

American Journal of Botany 66 (1979):
87–90.

PICKETT, CHARLES H., and W. DENNIS
CLARK. The Function of Extrafloral Nectaries in *Opuntia acanthocarpa* (Cacta-
ceae). *American Journal of Botany* 66
(1979): 618–25.

INOUYE, DAVID W., and ORLEY W.
TAYLOR, JR. A Temperate Region Plant–
Ant–Seed Predator System: Consequences
of Extra Floral Nectar Secretion by *Helian-
thella quinquenervis. Ecology* 60 (1979):
1–7.

LAINE, KARI J., and PEKKA NIEMELA. Influence of Ants on Survival of Mountain
Birches During an *Oporinia autumnata*
(Lep., Geometridae) Outbreak. *Oecologia*
47 (1980): 39–42.

TILMAN, DAVID. Cherries, Ants and Tent
Caterpillars: Timing of Nectar Production in Relation to Susceptibility of Caterpillars to Ant Predation. *Ecology* 59
(1978): 686–92.

CHAPTER 3

Self-Defense: Fleeing from the Wrath to Come

MACARTHUR, ROBERT H. *Geographical
Ecology.* New York: Harper and Row,
1972.

SCRIBER, J. MARK, and PAUL FEENY. Growth of Herbivorous Caterpillars in Relation to Feeding Specialization and to the Growth Form of Their Food Plants. *Ecology* 60 (1979): 829–50.

JONES, ROBERT E. Search Behavior: A Study of Three Caterpillar Species. *Behavior* 60 (1977): 237–58.

MEEUSE, BASTIAAN J.D. Dormancy and Germination of Seeds. University of Wisconsin Arboretum *Bulletin* 41 (1978): 8–12.

JANZEN, DANIEL. Seed Predation by Animals. *Annual Review of Ecology and Systematics* 2 (1971): 465–92.

———. Why Bamboos Wait So Long to Flower. *Annual Review of Ecology and Systematics* 7 (1976): 347–91.

LIGON, J. DAVID. Reproductive Interdependence of Piñon Jays and Piñon Pines. *Ecological Monographs* 48 (1978): 111–26.

OWEN, DENIS F. The Effect of a Consumer, *Phytomyza ilicis*, on Seasonal Leaf-Fall in the Holly, *Ilex aquifolium*. *Oikos* 31 (1978): 268–71.

———. Why Do Aphids Synthesize Melezitose? *Oikos* 31 (1978): 264–67.

BOUCHER, DOUGLAS H., and VICTORIA L. SORK. Early Drop of Nuts in Response to Insect Infestation. *Oikos* 33 (1979): 440–43.

CHAPTER 4
*Self-Defense: Repellents, Poisons,
and Contraceptives*

ESHLEMAN, ALAN. *Poison Plants*. Boston: Houghton Mifflin, 1977.

FRAENKEL, GOTTFRIED. The Raison d'Etre of Secondary Plant Substances. *Science* 129 (1959): 1466–70.

MAUGH, THOMAS H., II. Exploring Plant Resistance to Insects. *Science* 216 (1982): 722–23.

MORTON, JULIA. Notes on Economic Botany in Epidemiology. *Economic Botany* 32 (1978): 111–16.

FARENTINOS, R.C., P.J. CAPRETTA, R.E. KEPNER, and V.M. LITTLEFIELD. Selective Herbivory in Tassel-Eared Squirrels: Role of Monoterpenes in Ponderosa Pines Chosen as Feeding Trees. *Science* 213 (1981): 1273–75.

EDMUNDS, GEORGE F., and DONALD N. ALSTAD. Coevolution in Insect Herbivores and Conifers. *Science* 199 (1978): 941–45.

DORIA, JOHN J. Neem: The Tree Insects Hate. *Garden* July/August 1981: 8–11.

HAMILTON, W. GIBSON. *Our Edible Toadstools and Mushrooms and How to Distinguish Them*. New York: Harper and Brothers, 1895.

SWAIN, THEODORE. Plant–Animal Coevolution: A Synoptic View of the Paleozoic and Mesozoic. *Annual Proceedings of the Phytochemical Society*, vol. 15, 1978.

CATES, REX G. Host Plant Predictability and the Feeding Patterns of Monophagous, Oligophagous, and Phytophagous Insect Herbivores. *Oecologia* 48 (1981): 319–26.

FEENY, PAUL. Plant Apparency and Plant Defense. *Annual Proceedings of the Phytochemical Society*, vol. 10, 1976.

JACOBSON, MARTIN. Plants, Insects and Man —Their Interrelationships. *Economic Botany* 36 (1982): 346–54.

IVIE, G. WAYNE, DOUGLAS L. HOLT, and MARCELLUS C. IVEY. Natural Toxicants in Human Foods: Psoralens in Raw and Cooked Parsnip Root. *Science* 213 (1981): 909–10.

JONES, D.A., R.J. KEYMER, and W.M. ELLIS. Cyanogenesis in Plants and Animal Feeding. *Annual Proceedings of the Phytochemical Society*, vol. 15, 1978.

DIRZO, RODOLFO, and JOHN L. HARPER. Experimental Studies on Slug–Plant Interactions, III. *Journal of Ecology* 70 (1982): 101–17.

HAUKIOJA, ERKKI. On the Role of Plant Defences in the Fluctuation of Herbivore Populations. *Oikos* 35 (1980): 202–13.

BRYANT, JOHN P. Phytochemical Deterrence

of Snowshoe Hare Browsing by Adventitious Shoots of Four Alaskan Trees. *Science* 213 (1981): 889–90.

SCHULTZ, JACK C., and IAN T. BALDWIN. Oak Leaf Quality Declines in Response to Defoliation by Gypsy Moth Larvae. *Science* 217 (1982): 149–51.

SCHWEITZER DALE F. Effects of Foliage Age on Body Weight and Survival in Larvae of the Tribe Lithophanini (Lepidoptera: Noctuidae). *Oikos* 32 (1979): 403–8.

CHERETT, J.M. Chemical Aspects of Plant Attack by Leaf-Cutting Ants. *Annual Proceedings of the Phytochemical Society*, vol. 8, 1972.

RHOADES, DAVID F., and REX G. CATES. Toward a General Theory of Plant Antiherbivore Chemistry. *Annual Proceedings of the Phytochemical Society*, vol. 10, 1976.

McKEY, DOYLE, PETER G. WATERMAN, C.N. MBI, J.S. GARTLAN, and T.T. STRUHSAKER. Phenolic Content of Vegetation in Two African Rain Forests: Ecological Implications. *Science* 202 (1978): 61–64.

OATES, JOHN F., PETER G. WATERMAN, and GILLIAN M. CHOO. Food Selection by the South Indian Leaf-Monkey, *Presbytis johnii*, in Relation to Leaf Chemistry. *Oecologia* 45 (1980): 45–56.

BELOVSKY, GARY E. Food Plant Selection by a Generalist Herbivore: The Moose. *Ecology* 62 (1981): 1020–30.

WIGGLESWORTH, VINCENT BRIAN. *The Principles of Insect Physiology*, 7th ed. London: Chapman and Hall, 1972.

SLÁMA, K., and C.M. WILLIAMS. Juvenile Hormone Activity for the Bug *Pyrrhocoris apterus*. *Proceedings of the National Academy of Sciences U.S.A.*, vol. 54, 1965, 411–14.

BOWERS, WILLIAM S., T. OHTA, J.S. CLEERE, and P.A. MARSELLA. Discovery of Insect Anti-Juvenile Hormones in Plants. *Science* 193 (1976): 542–47.

BERGER, PATRICIA J., EDWARD H. SANDERS, PETE D. GARDNER, and NORMAN C. NEGUS. Phenolic Plant Compounds Functioning as Reproductive Inhibitors in *Microtus montanus*. *Science* 195 (1977): 575–77.

BERGER, PATRICIA J., NORMAN C. NEGUS, EDWARD H. SANDERS, and PETE D. GARDNER. Chemical Triggering of Reproduction in *Microtus montanus*. *Science* 214 (1981): 69–70.

LEOPOLD, A. STARKER, MICHAEL ERWIN, JOHN OH, and BRUCE BROWNING. Phytoestrogens: Adverse Effects on Reproduction in California Quail. *Science* 191 (1976): 98–100.

BAILEY, D. Plants and Medicinal Chemistry.

Education in Chemistry 14 (1977): 114–16 and 140–44.

ATSATT, PETER R., and DENNIS J. O'DOWD. Plant Defense Guilds. *Science* 193 1976): 24–29.

CHAPTER 5

Self-Defense: One Man's Meat Is Another Man's Poison

ERICKSON, JAMES M., and PAUL FEENY. Sinigrin: A Chemical Barrier to the Black Swallowtail Butterfly *Papilio polyxenes.* *Ecology* 55 (1974): 103–11.

JANZEN, DANIEL H., HARVEY B. JUSTER, and IRWIN E. LIENER. Insecticidal Action of the Phytohemagglutinin in Black Beans on a Bruchid Beetle. *Science* 192 (1976): 795–96.

KRIEGER, ROBERT I., PAUL FEENY, and C.F. WILKINSON. Detoxification Enzymes in the Guts of Caterpillars: An Evolutionary Answer to Plant Defenses? *Science* 172 (1971): 579–81.

BRATTSTEN, L.B., C.F. WILKINSON, and THOMAS EISNER. Herbivore–Plant Interactions: Mixed-Function Oxidases and Secondary Plant Substances. *Science* 196 (1977): 1349–52.

Urquhart, Frederick. *The Monarch Butterfly.* Toronto: University of Toronto Press, 1960.

BROWER, LINCOLN P. *Strategy for Survival.* 16-mm. color film, 30 minutes long, from Harper & Row Media, Hagerstown, Maryland, 1977.

————. Ecological Chemistry. *Scientific American* 220 (1969): 22–29.

BROWER, LINCOLN P., JANE VAN ZANDT BROWER, and JOSEPH M. CORVINO. Plant Poisons in a Terrestrial Food Chain. *Proceedings of the National Academy of Sciences of the U.S.A.,* vol. 57, 1967, 893–98.

BROWER, LINCOLN P., and SUSAN C. GLAZIER. Localization of Heart Poisons in the Monarch Butterfly. *Science* 188 (1975): 19–25.

PASTEELS, JACQUES M., and DÉSIRÉ DALOZE. Cardiac Glycosides in the Defensive Secretion of Chrysomelid Beetles: Evidence for Their Production by the Insects. *Science* 197 (1977): 70–72.

EDMUNDS, M. *Defence in Animals.* London: Longmans, 1974.

BERENBAUM, MAY R. Patterns of Furanocoumarin Production and Insect Herbivory in a Popultaion of Wild Parsnip. *Oecologia* 49 (1981): 236–44.

JÄRVI, TORBJORN, BIRGITTA SILLEN-TULLBERG, and CHRISTER WIKLUND. The Cost of Being Aposematic: An Experimental Study of Predation on Larvae of

Papilio machaon by the Great Tit *Parus major*. *Oikos* 36 (1981): 267–72.

ROTHSCHILD, MIRIAM, ROBIN APLIN, JOHN BAKER, and NEVILLE MARSH. Toxicity Induced in the Tobacco Hornworm (*Manduca sexta* L.) (Sphingidae, Lepidoptera). *Nature* 280 (1979): 487–88.

WILSON, E.O. *Sociobiology.* Cambridge, Massachusetts: Harvard University Press, 1975.

KROPOTKIN, PETER. *Mutual Aid.* New York: Alfred Knopf, 1921.

ROTHSCHILD, MIRIAM, and D.N. KELLETT. Reactions of Various Predators to Insects Storing Heart Poisons (Cardiac Glycosides) in Their Tissues. *Journal of Entomology* (A) 46 (1972): 103–10.

CALVERT, WILLIAM H., LEE E. HEDRICK, and LINCOLN P. BROWER. Mortality of the Monarch Butterfly (*Danaus plexippus* L.): Avian Predation at Five Overwintering Sites in Mexico. *Science* 204 (1979): 847–51.

FINK, LINDA S., and LINCOLN P. BROWER. Birds Can Overcome the Cardenolide Defence of Monarch Butterflies in Mexico. *Nature* 291 (1981): 67–70.

THORPE, WILLIAM H. *Learning and Instinct in Animals.* Cambridge, Massachusetts: Harvard University Press, 1956.

GARCIA, JOHN, WALTER G. HANKINS, and KENNETH W. RUSINIAK. Behavioral Regulation of the *Milieu Interne* in Man and Rat. *Science* 185 (1974): 824–31.

ROZIN, PAUL, and JAMES W. KALAT. Specific Hungers and Poison Avoidance as Adaptive Specializations of Learning. *Psychological Review* 78 (1971): 459–86.

GELPERIN, ALAN. Rapid Food-Aversion Learning by a Terrestrial Mollusk. *Science* 189 (1975): 567–70.

SMITH, D.A.S. The Significance of Beak Marks on the Wings of an Aposematic, Distasteful and Polymorphic Butterfly. *Nature* 281 (1979): 215–16.

CHAPTER 6
Lilies of the Field: Flowers and Cross-Pollination

FAEGRI, KARL, and LEENDERT VAN DER PIJL. *The Principles of Pollination Ecology*, 2nd ed. Oxford: Pergamon Press, 1971.

TAKHTAJIAN, A.L. *Origins of Angiospermous Plants*, translated by O.H. Gankin and edited by G.L. Stebbins. Washington, D.C.: American Institute of Biological Sciences, 1958.

MEEUSE, BASTIAAN J.D. *The Story of Pollination*. New York: Ronald Press, 1961.

MULCAHY, DAVID L. The Rise of the Angiosperms: A Genecological Factor. *Science* 206 (1979): 20–23.

GRANT, VERNE. The Fertilization of Flowers. *Scientific American* 184 (1951): 52–56.

BOGGS, CAROL L., JOHN T. SMILEY, and LAWRENCE E. GILBERT. Patterns of Pollen Exploitation by *Heliconius* Butterflies. *Oecologia* 48 (1981): 284–89.

THIEN, LEONARD B. Floral Biology of *Magnolia. American Journal of Botany* 61 (1974): 1037–45.

————. Patterns of Pollination in the Primitive Angiosperms. *Biotropica* 12 (1980): 1–13.

PIJL, LEENDERT VAN DER, and CALAWAY H. DODSON. *Orchid Flowers: Their Pollination and Evolution.* Coral Gables, Florida: University of Miami Press, 1966.

RICKSON, FRED R. Ultrastructural Development of the Beetle Food Tissue of *Calycanthus* Flowers. *American Journal of Botany* 66 (1979): 80–86.

MEEUSE, BASTIAAN J.D. The Voodoo Lily, *Scientific American* 215 (1966): 80–88.

IVRI, YARIV, and AMOTS DAFNI. The Pollination Ecology of *Epipactis consimilis* Don (Orchidaceae) in Israel. *New Phytologist* 79 (1977): 173–77.

FREE, JOHN. *Insect Pollination of Crops.* New York: Academic Press, 1970.

YOUNG, ALLEN M. The Confection Connection. *Garden* 7 (1983): 21–32.

ILSE, DORA. New Observations on Response to Colors in Egg-Laying Butterflies. *Nature* 140 (1937): 544–45.

SWIHART, CHRISTINE A., and S.L. SWIHART. Color Selection and Learned Feeding Preferences in the Butterfly, *Heliconius charitonius* Linn. *Animal Behavior* 18 (1970): 60–74.

CRUDEN, R.W., and SHARON M. HERMANN-PARKER. Butterfly Pollination of *Caesalpinia pulcherrima*, with observations on a Psychophilous Syndrome. *Journal of Ecology* 67 (1979): 155–68.

WIKLUND, CHRISTER, TORSTEN ERIKSSON, and HANS LUNDBERG. The Wood White Butterfly *Leptidea sinapsis* and Its Nectar Plants: *A Case of Mutualism or Parasitism? Oikos* 33 (1979): 358–62.

PROCTOR, MICHAEL, and PETER YEO. *The Pollination of Flowers*. New York: Taplinger Publishing Company, 1972.

HICKMAN, JAMES C. Pollination By Ants: A Low-Energy System. *Science* 184 (1974): 1290–92.

CHAPTER 7
*How Doth the Little Busy Bee: Honeybees,
Bumblebees, and Wasps as Pollinators*

VON FRISCH, KARL. *The Dance Language and Orientation of Bees.* Cambridge, Massachusetts: Harvard University Press, 1967.

LINDAUER, MARTIN. *Communication Among Social Bees.* Cambridge, Massachusetts: Harvard University Press, 1961.

HEINRICH, BERND. *Bumblebee Economics.* Cambridge, Massachusetts: Harvard University Press, 1979.

WILSON, EDWARD O. *The Insect Societies.* Cambridge, Massachusetts: Harvard University Press, 1971.

WHITHAM, THOMAS G. Coevolution of Foraging in *Bombus* and Nectar Dispensing in *Chilopsis*: A Last Dreg Theory. *Science* 197 (1977): 593–96.

BRINK, DONALD E. Reproduction and Variation in *Aconitum columbianum* (Ranunculaceae) with Emphasis on California Populations. *American Journal of Botany* 67 (1980): 263–73.

BRINK, DONALD E., and J.M.J. DE WET. Interpopulation Variation in Nectar Production in *Aconitum columbianum* (Ranunculaceae). *Oecologia* 47 (1980): 160–63.

MORSE, DOUGLASS H. Modification of Bumblebee Foraging: the Effect of Milkweed Pollinia. *Ecology* 62 (1981): 89–97.

WYATT, ROBERT. The Reproduction Biology of *Asclepias tuberosa*: I. Flower Number, Arrangement, and Fruit-Set. *New Phytologist* 85 (1980): 119–31.

MEEUSE, BASTIAAN J.D. Flower Constancy: Experiment No. 50. *Laboratory Exercises* for Botany 475, University of Washington, n.d.

VAN DER PIJL, LEENDERT, and CALAWAY H. DODSON. *Orchid Flowers: Their Pollination and Evolution*. Coral Gables, Florida: University of Miami Press, 1966.

PROCTOR, MICHAEL, and PETER YEO. *The Pollination of Flowers*. New York: Taplinger, 1972.

FREE, JOHN B. *Insect Pollination of Crops*. New York: Academic Press, 1970.

CHAPTER 8
Hummers, Creepers, and Fly-by-Nights: Warm-Blooded Pollinators

VAN DER PIJL, LEENDERT. Ecological Aspects of Flower Evolution, II. Zoophilous Flower Classes. *Evolution* 15 (1961): 44–59.

GRANT, KAREN A., and VERNE GRANT. *Hummingbirds and Their Flowers*. New York: Columbia University Press, 1968.

BROWN, JAMES H., and ASTRID KODRIC-

BROWN. Convergence, Competition and Mimicry in a Temperate Community of Hummingbird-Pollinated Flowers. *Ecology* 60 (1979): 1022–35.

WELTY, JOEL C. *The Life of Birds*, 2nd ed. Philadelphia: W.B. Saunders, 1975.

PIANKA, ERIC R. *Evolutionary Ecology*. New York: Harper & Row, 1978.

ROUBIK, DAVID W. The Ecological Impact of Nectar-Robbing Bees and Pollinating Hummingbirds on a Tropical Shrub. *Ecology* 63 (1982): 354–60.

MILLER, RUSSELL B. The Pollination Ecology of *Aquilegia elegantula* and *A. caerulea* (Ranunculaceae) in Colorado. *American Journal of Botany* 65 (1978). 406–14.

WASER, NICKOLAS M. Competition for Hummingbird Pollination and Sequential Flowering in Two Colorado Wildflowers. *Ecology* 59 (1978): 934–44.

STILES, F. GARY. Coadapted Competitors: The Flowering Seasons of Hummingbird-Pollinated Plants in a Tropical Forest. *Science* 198 (1977): 1177–78.

FEINSINGER, PETER. Ecological Interaction Between Plants and Hummingbirds in a Successional Tropical Community. *Ecological Monographs* 48 (1978): 269–87.

FEINSINGER, PETER, JAMES A. WOLFE, and LEE ANN SWARM. Island Ecology: Reduced Hummingbird Diversity and the

Pollination of Plants, Trinidad and Tobago, West Indies. *Ecology* 63 (1982): 494–506.

LINHART, YAN B., and PETER FEINSINGER. Plant–Hummingbird Interactions: Effects of Island Size and Degree of Specialization on Pollination. *Journal of Ecology* 68 (1980): 745–60.

GILL, FRANK B., and LARRY L. WOLF. Nonrandom Foraging by Sunbirds in a Patchy Environment. *Ecology* 58 (1977): 1284–96.

MEEUSE, BASTIAAN J.D. *The Story of Pollination.* New York: Ronald Press, 1961.

FAEGRI, KARL, and LEENDERT VAN DER PIJL. *The Principles of Pollination Ecology*, 2nd ed. Oxford: Pergamon Press, 1971.

LACK, ANDREW. The Ecology of the Flowers of the Savanna Tree *Maranthes polyandra* and Their Visitors, with Particular Reference to Bats. *Journal of Ecology* 66 (1978): 287–95.

NOVICK, ALVIN, and NINA LEEN. *The World of Bats.* New York: Holt, Rinehart and Winston, 1969.

GORMAN, JAMES. Beauty and the Bats. *Discover* 1981: 30–33.

HOWELL, DONNA J. Flock Foraging in Nectar-Feeding Bats: Advantages to the Bat and to the Host Plants. *American Naturalist* 114 (1979): 23–49.

BERNHARDT, PETER. Pollinating Possums. *Garden* (1981): 22–25.

SUSSMAN, ROBERT W., and PETER H. RAVEN. Pollination by Lemurs and Marsupials: An Archaic Coevolutionary System. *Science* 200 (1978): 731–36.

HARGREAVES, DOROTHY and BOB. Tropical Trees. Portland, Oregon: Hargreaves Industrial, 1965.

WIENS, DELBERT, and JOHN P. ROURKE. Rodent Pollination in Southern African *Protea spp. Nature* 276 (1978): 71–73.

CHAPTER 9

*Children, Go Where I Send Thee:
Animal-Aided Plant Dispersal*

VAN DER PIJL, LEENDERT. *Principles of Dispersal in Higher Plants.* Berlin: Springer-Verlag, 1969.

LEMEN, CLIFF. Elm Trees and Elm Leaf Beetles: Patterns of Herbivory. *Oikos* 36 (1981): 65–67.

REGAL, PHILIP J. Ecology and Evolution of Flowering Plant Dominance. *Science* 196 (1977): 622–29.

JANZEN, DANIEL. Fruits for Famished Mammoths. *Garden* (1982): 12–32.

VANDER WALL, STEPHEN B., and RUSSEL P. BALDA. Coadaptations of the Clark's Nutcracker and the Piñon Pine for Efficient Seed Harvest and Dispersal. *Ecological Monographs* 47 (1977): 89–111.

LIGON, J. DAVID. Reproductive Interdependence of Piñon Jays and Piñon Pines. *Ecological Monographs* 48 (1978): 111–26.

MEEUSE, BASTIAAN J.D. Myrmecochory, or the Dispersal of Seeds by Ants. *Proceedings of the Washington State Entomological Society*, vol. 40, 1978, 541–48.

HEITHAUS, E. RAYMOND, DAVID C. CULVER, and ANDREW J. BEATTIE. Models of Some Ant–Plant Mutualisms. *American Naturalist* 116 (1980): 347–61.

HEITHAUS, E. RAYMOND. Seed Predation by Rodents on Three Ant-Dispersed Plants. *Ecology* 62 (1981): 136–45.

CULVER, DAVID C., and ANDREW J. BEATTIE. Myrmecochory in *Viola*: Dynamics of Seed–Ant Interactions in Some West Virginia Species. *Journal of Ecology* 66 (1978): 53–72.

———. The Fate of *Viola* Seeds Dispersed by Ants. *American Journal of Botany* 67 (1980): 710–14.

BROWN, JAMES H., and DIANE W. DAVIDSON. Competition between Seed-Eating Rodents and Ants in Desert Ecosystems. 196 (1977): 880–82.

DAVIDSON, DIANE W., and S.R. MORTON. Competition for Dispersal in Ant-Dispersed Plants. *Science* 213 (1981): 1259–61.

BEATTIE, ANDREW J., and DAVID C. CULVER. The Guild of Myrmecochores in the Herbaceous Flora of West Virginia Forests. *Ecology* 62 (1981): 107–15.

BERG, ROLF Y. Dispersal Ecology of *Vancouveria* (Berberidaceae). *American Journal of Botany* 59 (1972): 109–22.

RUEHLE, JOHN L., and DONALD H. MARX. Fiber, Food, Fuel, and Fungal Symbionts. *Science* 206 (1979): 419–22.

MASER, CHRIS, JAMES M. TRAPPE, AND DOUGLAS C. URE. Implications of Small Mammal Mycophagy to the Management of Western Coniferous Forests. *Transactions of the 43rd North American Wildlife and Natural Resources Conference,* 1978, 78–88.

TRAPPE, JAMES M., and CHRIS MASER. Germination of Spores of *Glomus macrocarpus* (Endogonaceae) After Passage Through a Rodent Digestive Tract. *Mycologia* 67 (1976): 433–36.

————. Ectomycorrhizal Fungi: Interactions of Mushrooms and Truffles with Beasts and Trees, in *Mushrooms and Man*, Forest Service, U.S.D.A., 1977.

MASER, CHRIS, JAMES M. TRAPPE, and RONALD A. NUSSBAUM. Fungal–Small Mammal Interrelationships with Emphasis on Oregon Coniferous Forests. *Ecology* 59 (1978): 799–809.

BERGSTROM, DOROTHY. Small Mammals

Traffic in Truffles. *Forestry Research West* (1978): 1–4.

SWAIN, ROGER. Science and the Gardener: Acocadoes. *Horticulture* 61 (1983): 10–13.

MAUGH, THOMAS H., II. A Fish in the Bush Is Worth . . . *Science* 211 (1981): 1151.

KUIJT, JOB. *The Biology of Parasitic Flowering Plants.* Berkeley: University of California Press, 1969.

GLYPHIS, J.P., S.T. MILTON, and W.R. SIEGFRIED. Dispersal of *Acacia cyclops* by Birds. *Oecologia* 48 (1981): 138–41.

TEMPLE, STANLEY A. Plant–Animal Mutualism: Coevolution with Dodo Leads to Near Extinction of Plant. *Science* 197 (1977): 885–86.

KUMMER, HANS. *Social Organization of Hamadryas Baboons: A Field Study.* Chicago: University of Chicago Press, 1968.

MORRISON, DOUGLAS W. Foraging Ecology and Energetics of the Frugivorous Bat *Artibeus jamaicensis. Ecology* 59 (1978): 716–23.

HEITHAUS, E. RAYMOND, and THEODORE H. FLEMING. Foraging Movements of a Frugivorous Bat, *Carollia perspicillata* (Phyllostomatidae). *Ecological Monographs* 48 (1978): 127–43.

FLEMING, THEODORE H., E. RAYMOND HEITHAUS, and WILLIAM B. SAWYER. An Experimental Analysis of the Food

Location Behavior of Frugivorous Bats. *Ecology* 58 (1977): 619–27.

MORRISON, DOUGLAS W. Lunar Phobia in a Neotropical Fruit Bat, *Artibeus jamaicensis* (Chiroptera: Phyllostomidae). *Animal Behavior* 26 (1978): 852–55.

————. Efficiency of Food Utilization by Fruit Bats. *Oecologia* 45 (1980): 270–73.

CHAPTER 10
Whither Thou Goest: Long-Term Animal–Plant Symbioses

HENRY, S.M., ed. *Symbiosis.* New York: Academic Press, 1967.

BATRA, LEKH R. Ambrosia Fungi: Extent of Specificity to Ambrosia Beetles. *Science* 153 (1966): 193–95.

WILSON, EDWARD O. *The Insect Societies.* Cambridge, Massachusetts: Harvard University Press, 1971.

MARTIN, MICHAEL M., and JOAN S. MARTIN. Cellulose Digestion in the Midgut of the Fungus-Growing Termite *Macrotermes natalensis*: The Role of Acquired Digestive Enzymes. *Science* 199 (1978): 1453–55.

BELT, THOMAS. *The Naturalist in Nicaragua.* New York: E.P. Dutton, 1911.

COOKE, RODERICK. *The Biology of Symbiotic*

Fungi. London: John Wiley and Sons, 1977.

RAVEN, PETER H., RAY F. EVERT, and HELENA CURTIS. *Biology of Plants*, 3rd ed. New York: Worth, 1981.

MUSCATINE, LEONARD, and H.M. LENHOFF. Symbiosis of *Hydra* and Algae, II. Effects of Limited Food and Starvation on Growth of Symbiotic and Aposymbiotic *Hydra*. *Biological Bulletin* 129 (1965): 316–28.

YONGE, C.M. Giant Clams. *Scientific American* 232 (1975): 96–105.

FITT, WILLIAM, and ROBERT K. TRENCH. Spawning, Development, and Acquisition of Zooxanthellae by *Tridacna squamosa* (Mollusca, Bivalvia). *Biological Bulletin* 161 (1981): 231–35.

SMITH, DAVID, LEONARD MUSCATINE, and DAVID LEWIS. Carbohydrate Movement from Autotrophs to Heterotrophs in Parasitic and Mutualistic Symbiosis. *Biological Reviews of the Cambridge Philosophical Society* 44 (1969): 17–90.

MUSCATINE, LEONARD, and JAMES W. PORTER. Reef Corals and Mutualistic Symbioses Adapted to Nutrient-Poor Environments. *Bioscience* 27 (1977): 454–60.

RAHAT, M., and ORIT ADAR. Effect of Symbiotic Zooxanthellae and Temperature on Budding and Strobilation in *Casseiopeia andromeda*. *Biological Bulletin* 159 (1980): 394–401.

NEISS, DORTE, WERNER REISSER, and WOLFGANG WEISSNER. The Role of Endosymbiotic Algae in Photoaccumulation of Green *Paramecium bursaria*. *Planta* 152 (1981): 268–71.

BUCHSBAUM, VICKI PEARSE. Modification of Sea Anemone Behavior by Symbiotic Zooxanthellae: Phototaxis. *Biological Bulletin* 147 (1974): 630–40.

GLADFELTER, WILLIAM B. Sea Anemone with Zooxanthellae: Simultaneous Contraction and Expansion in Response to Changing Light Intensity. *Science* 175 (1975): 570–71.

PHIPPS, D.W., JR., and ROOSEVELT L. PARDY. Host Enhancement of Symbiont Photosynthesis in the *Hydra*–Algae Symbiosis. *Biological Bulletin* 162 (1982): 83–94.

PARK, HELEN D., C.L. GREENBLATT, C.F.T. MATTERN, and C.R. MERRILL. Some Relationships Between *Chlorhydra*, Its Symbionts and Some Other Chlorophyllous Forms. *Journal of Experimental Zoology* 164 (1967): 141–62.

PARDY, ROOSEVELT, and LEONARD MUSCATINE. Recognition of Symbiotic Algae by *Hydra viridis*. A Quantitative Study of the Uptake of Living Algae by Aposymbiotic

H. viridis. Biological Bulletin 145 (1973): 565–79.

KINZIE, ROBERT A., III, and GEORGENE S. CHEE. The Effect of Different Zooxanthellae on the Growth of Experimentally Reinfected Hosts. *Biological Bulletin* 156 (1979): 315–27.

MARGULIS, LYNN. *Early Life.* Boston: Science Books International, 1982.

JENNINGS, D. H., and D.L. LEE, eds. *Symbiosis.* Symposia of the Society for Experimental Biology XXIX. Cambridge, England: Cambridge University Press, 1975.

CHENG, THOMAS, ed. *Aspects of the Biology of Symbiosis.* Baltimore: University Park Press, 1971.

Index

Index

(208)